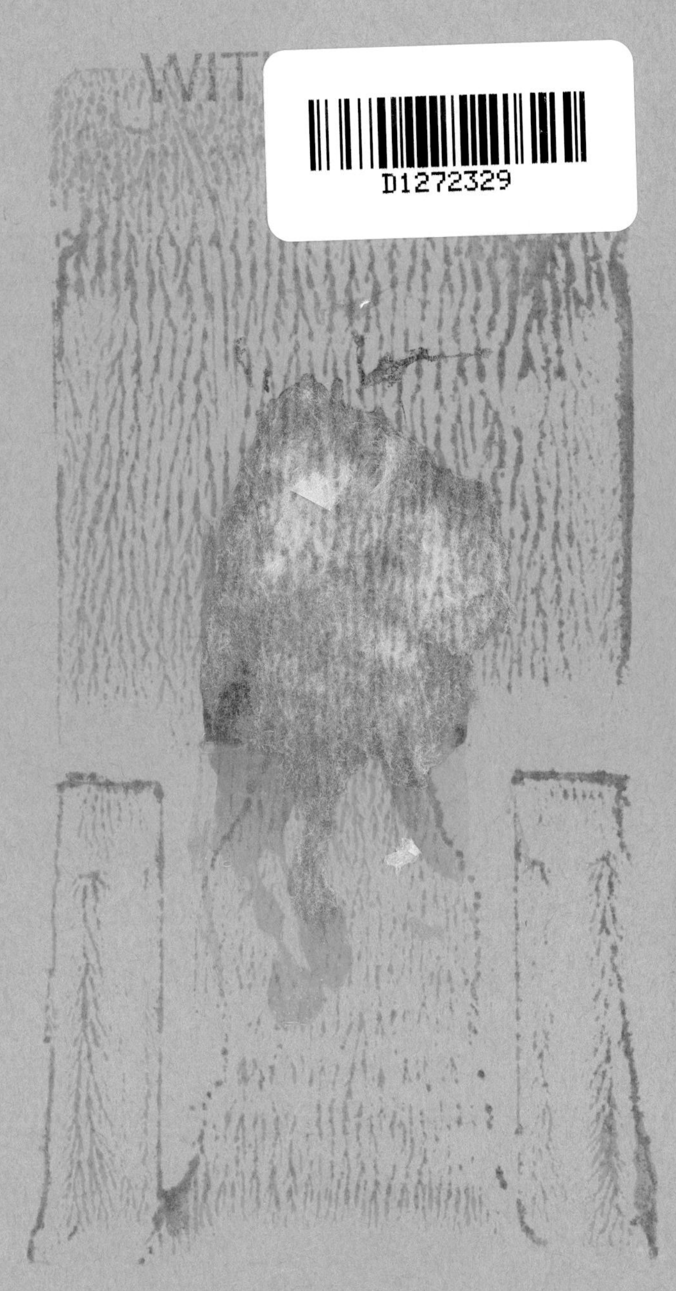

THE ILLUSTRATED FLORA OF ILLINOIS

The Illustrated Flora of Illinois

ROBERT H. MOHLENBROCK, General Editor

THE ILLUSTRATED FLORA OF ILLINOIS

FERNS

Robert H. Mohlenbrock

SOUTHERN ILLINOIS UNIVERSITY PRESS
Carbondale and Edwardsville
FEFFER & SIMONS, INC.
London and Amsterdam

Library of Congress Catalog Number 65–16533
Printed in the United States of America
Designed by Andor Braun

This book is dedicated to

Miss E. Esther Smith,

former biology teacher from Murphysboro (Ill.)

Township High School,

who first instilled within the author

a love for nature.

FOREWORD

After having worked with various aspects of the Illinois flora for over a decade, I came to the realization that not a great amount of information was known about all the Illinois flora, and that which was known was of a nature which was virtually useless to the average person wanting to know about the plants of this state. Thus the idea was conceived to attempt something that had never before been accomplished for any one of the United States—a multivolumed flora of the state of Illinois, to cover every group of plants, from algae and fungi through flowering plants. In addition to an account with keys of every plant known to occur in Illinois, there would be provided illustrations showing the diagnostic characters of each species.

An advisory board was set up in 1964 to screen, criticize, and make suggestions for each volume of The Illustrated Flora of Illinois during its preparation. The board is composed of taxonomists eminent in their area of specialty—Dr. Gerald W. Prescott, Michigan State University (algae), Dr. Constantine J. Alexopoulos, University of Texas (fungi), Dr. Aaron J. Sharp, University of Tennessee (bryophytes), Dr. Rolla M. Tryon, Jr., The Gray Herbarium (ferns), and Dr. Robert F. Thorne, Rancho Santa Ana Botanical Garden (flowering plants).

This author is editor of the series, and will prepare many of the volumes. Specialists in various groups will be asked to prepare the sections of their special interest.

There is no definite sequence for publication of The Illustrated Flora of Illinois. Rather, volumes will appear as they are completed.

Robert H. Mohlenbrock

Southern Illinois University

CONTENTS

ILLUSTRATIONS

The Illustrated Flora of Illinois

FERNS

County Map of Illinois

Introduction

The nomenclature followed in this volume is based largely on that of E. T. Wherry (1961) in *The Fern Guide.* Synonyms, with complete author citation, which have applied to species in the northeastern United States are given under each species. A description, based primarily on Illinois material, is provided for each species. The description, while not intended to be complete, covers the important features of the species.

The common name, or names, is the one used locally in Illinois. The habitat designation is not always the habitat throughout the range of the species, but only for it in Illinois. The over-all range for each species is given from the northeastern to the northwestern extremities, south to the southwestern limit, then eastward to the southeastern limit. The range has been compiled from various sources, including examination of herbarium material. A general statement is given concerning the range of each species in Illinois. Dot maps showing county distribution of each fern in Illinois are provided. Each dot represents a voucher specimen deposited in some herbarium. There has been no attempt to locate each dot with reference to the actual locality within each county.

The distribution has been compiled from field study as well as herbarium study. Herbaria from which specimens have been studied are the Chicago Museum of Natural History, Eastern Illinois University, the Gray Herbarium of Harvard University, Illinois Natural History Survey, Illinois State Museum, Missouri Botanical Garden, New York Botanical Garden, Southern Illinois University, the United States National Herbarium, the University of Illinois, and Western Illinois University. In addition, a few private collections have been examined.

Each species is illustrated, showing the habit as well as some of the distinguishing features in detail. Mrs. Miriam Meyer has prepared all of the illustrations.

There are several persons to whom the author is indebted for assistance in this study. Dr. Rolla M. Tryon of the Gray Herbarium has read and commented on the entire manuscript. Dr. Warren H. Wagner, University of Michigan, has read the manuscript to *Dryopteris* and *Asplenium* and supplied very valuable comments. Dr. Richard Hauke, University of Rhode Island, has done the same for *Equisetum*. Mrs. Alice Tryon has made valuable suggestions concerning *Pellaea*. For courtesies extended in their respective herbaria, the author is indebted to Dr. Robert A. Evers, Illinois Natural History Survey; Dr. G. Neville Jones, University of Illinois; Dr. Glen S. Winterringer, Illinois State Museum; Dr. Arthur Cronquist, New York Botanical Garden; Dr. Jason Swallen and Mr. Conrad V. Morton, United States National Herbarium; and to the herbarium and library staff of the Missouri Botanical Garden. Southern Illinois University provided time and space for the preparation of this work. The Graduate School of Southern Illinois University provided salary for the illustrator, while the Mississippi Valley Investigations and its late director, Dr. Charles C. Colby, furnished funds for some of the field work.

HISTORY OF FERN COLLECTING IN ILLINOIS

Perhaps no other group of plants attracts more interest among amateur botanists than ferns. This interest by so many persons, along with the relatively limited number of ferns in Illinois, makes the distribution better known for ferns than for any other group of plants.

As early as 1846, when S. B. Mead published one of the first lists of Illinois plants, sixteen species were already known from the state. Activity among plant hunters in Illinois was vigorous during the last half of the nineteenth century, with botanists such as Mead, Lapham, Vasey, Patterson, Brendel, Eggert, and Hill the most noteworthy. Their contributions enabled Patterson in 1876 and Brendel in 1887 to record fifty and forty-eight species, respectively, of ferns and fern-allies in Illinois. Between 1887 and 1923, through the efforts of Hill, Gleason, Steagall, and others, twelve additional species were found in Illinois. With the discovery by Steagall in 1923 of the filmy fern, the total number of ferns and fern-allies from Illinois stood at sixty-two.

Then began an era where fern collecting was at a virtual standstill. For nearly three decades, beginning in 1923, only a single species of fern was added to the Illinois flora—that being *Woodwardia virginica*, collected in 1944 by Fuller and Jones. Swayne found *Lycopodium flabelliforme* in Pope County in 1949. In 1952, Hooks made the original discovery of *Lycopodium dendroideum* in Illinois. In 1953, Voigt discovered *Isoetes butleri* and, one year later, Mohlenbrock found *Asplenium bradleyi.*

Thus, when Winterringer and Evers published their distribution of Illinois plants in 1960, sixty-seven species of ferns and fern-allies had been recorded from Illinois. During the research for the preparation of this volume thirteen additional species have been discovered, in addition to one previously unknown variety. Of these fourteen new taxa, five have been elevated from a lower rank, four represent previously overlooked or misidentified Illinois specimens, while five resulted from intensive field work. A total of eighty-one species and six lesser taxa is treated in this volume.

FERNS AND FERN-ALLIES

The plants treated in this volume are known popularly as ferns and fern-allies. They make up the old classification category of Division Pteridophyta, the vascular cryptogams—those plants with true vascular or conducting structures, but without flowers. More modern systems of classification tend to have these plants divided into three major groups of equal rank—the horsetails in one, the clubmosses and quillworts in another, and the remaining "ferns" in still another. Only plants belonging to the Filicineae of Christensen are considered "true ferns," the others known as "fern-allies." This latter term particularly applies to horsetails, clubmosses, and quillworts.

During the life history of all these plants, the conspicuous plant is the sporophyte, or spore-producing plant. Spores are usually all alike in a given species, but in *Selaginella, Isoetes, Marsilea,* and *Azolla* spores are of two kinds, termed microspores and megaspores. The spores germinate into a variable-shaped generation, the gametophyte, which produces the male and female gametes. In most genera, the gametophyte is green and rather heart-shaped or variously lobed, although in *Ophioglossum, Botrychium,* and some species of *Lycopodium,* the gametophyte is subterranean and therefore non-green.

Spores are borne in sporangia (*Fig. 1*). In most of the true ferns, the sporangia are arranged in distinct clusters, called sori (*Fig. 2*), on the lower surface of the blade. Occasionally the sori are formed along the margin of the blade, and may be continuous (*Fig. 3*) or interrupted (*Fig. 4*). Often a membranous indusium covers or partly covers the sorus (*Fig. 5*). In *Matteuccia* and *Onoclea,* the segments that bear the sori are much reduced so that the fertile leaves are quite different from the sterile ones (*Figs. 6 and 7*). In *Ophioglossum* and *Botrychium*, the sporangia are borne in two rows on terminal, unbranched (*Fig. 8*) or branched (*Fig. 9*) spikes, respectively. In *Osmunda*,

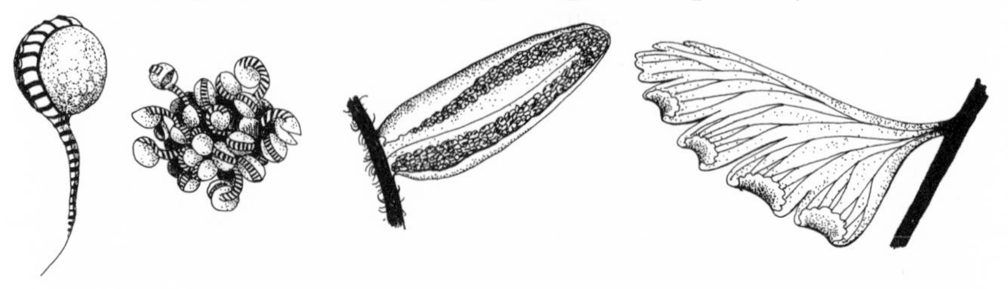

(Left to right) 1. Sporangium. 2. Sorus. 3. Continuous sori. 4. Interrupted sori.

(Left to right) 5. Sorus covered by indusium. 6. Fertile leaves (*Onoclea*). 7. Fertile leaves (*Matteuccia*). 8. Sporangia on unbranched spike (*Ophioglossum*). 9. Sporangia on branched spike (*Botrychium*).

the sporangia are borne in more than two rows, or irregularly (*Fig. 10*). *Trichomanes* has the sporangia at the base of a bristle-like projection from the leaf-margin. The sporangia in *Marsilea* and *Azolla* are enclosed in sporocarps. In *Marsilea*, the sporocarps are long-stalked and attached to the rhizome (*Fig. 11*), while in *Azolla*, they are sessile and produced along the filiform axis (*Fig. 12*). *Isoetes* has its sporangia at the base of the grass-like leaves (*Fig. 13*). The sporangia are borne in terminal cones (strobili) in *Equisetum, Selaginella,* and in some *Lycopodium* species (*Fig. 14*), although in other species of *Lycopodium,* the sporangia are borne in the axil of the leaf.

(Left to right) *10.* Sporangia borne irregularly (*Osmunda*). *11.* Long-stalked sporocarps attached to the rhizome (*Marsilea*). *12.* Sessile sporocarps along the filiform axis (*Azolla*). *13.* Sporangia at the base of grass-like leaves (*Isoetes*). *14.* Sporangia borne in terminal cones (*Equisetum, Selaginella,* and some *Lycopodium*).

The terminology applied to the various degrees of leaf-cutting must be thoroughly understood before one can become efficient at utilizing the keys. If the blade is lobed, but the cleft does not extend to the rachis so that the segment is sessile or stalked, the blade is pinnatifid (*Fig. 15*). If divisions extend to the rachis, then the blade is once-pinnate, and each division is called a pinna (*Fig. 16*). Each pinna may, in turn, be cleft only partway to the center, or it may be completely divided into distinct segments. In the first situation, the blade is pinnate-pinnatifid (*Fig. 17*); in the second, the blade is bipinnate (*Fig. 18*). These ultimate segments may again be partially cleft, bipinnate-pinnatifid (*Fig. 19*), or completely divided, tripinnate (*Fig. 20*).

The leaf is that part of the fern which includes not only the expanded blade but also the petiole. When measurements are given in this volume for the leaf, they are from the base of the petiole to the tip of the blade. The petiole extends from the base of the leaf to that point where the blade begins. The continuation of the petiole beyond that point is called the rachis.

(Above, left to right) *15*. Pinnatifid blade. *16*. Once-pinnate blade. *17*. Pinnate-pinnatifid blade. *18*. Bipinnate blade.

(Below, left to right) *19*. Bipinnate-pinnatifid blade. *20*. Tripinnate blade.

FERN HABITATS IN ILLINOIS

Limestone Outcroppings

Limestone outcroppings are most common along the major waterways in Illinois. As one proceeds south along the Mississippi River, for example, from JoDaviess County to Cairo, one may observe a more or less continuous range of limestone cliffs. These cliffs may come to the water's edge, or they may be a distance of several miles from the river. The same situation obtains along the Ohio River and, to some extent, along the Illinois River. The ferns which inhabit these limestone cliffs are usually confined to calcareous situations; one species—the walking fern (*Asplenium rhizophyllum*)—seems equally adapted to sandstone rocks. *Pellaea glabella, Asplenium rhizophyllum,* and *Cystopteris bulbifera* may be found on limestone throughout the state; *Cryptogramma stelleri* is restricted to northern Illinois; *Ophioglossum engelmannii, Pellaea atropurpurea,* and *Asplenium resiliens* are only in the southern half of the state. *Cheilanthes feei,* which is predominantly southern, is known also from two northern counties.

Sandstone Outcroppings

Sandstone outcroppings are more abundant in Illinois than are limestone outcroppings. They form the extensive Shawnee Hills, extending across extreme southern Illinois from near the Mississippi River to near the Ohio River; they are conspicuous in northern Illinois, for example, in LaSalle and Ogle counties. There is generally a distinct fern community on the exposed sandstone rocks or cliffs and on the heavily shaded rocks and cliffs. Only *Woodsia obtusa* seems equally at home under both extremes of moisture availability.

EXPOSED SANDSTONE The sandstone bluff tops regularly support the growth of nine species of ferns and fern-allies. Only *Woodsia ilvensis* is confined to the northern counties, while *Woodsia obtusa* ranges throughout the state. *Lycopodium flabelliforme* is known naturally from one northern and one southern county. The remaining six species are confined to the Shawnee Hills of southern Illinois. These are *Cheilanthes lanosa, Asplenium pinnatifidum, Asplenium × trudellii, Asplenium × kentuckiense, Asplenium bradleyi,* and *Isoetes butleri.* This last-named species occurs only in moist depressions in the sandstone bluff tops.

SHADED SANDSTONE Except for the moist woodlands, the shaded sandstone provides the habitat for the most species of ferns and fern-allies in Illinois. *Lycopodium dendroideum, Dryopteris × triploidea,* and *Dryopteris × boottii,* each known only from a single station, so far are restricted to northern Illinois; *Trichomanes boschianum, Dennstaedtia punctilobula, Polypodium polypodioides* var. *michauxianum*, *Asplenium* × *ebenoides*, and *Asplenium trichomanes* are restricted to southern Illinois. In addition, the three species of *Osmunda* in Illinois, while occupying swampy areas and lowland woods in the northern half of the state, are usually confined to moist sandstone ledges in the extreme southern counties. The majority of the moist sandstone species may be found in both northern and southern counties. These are *Lycopodium porophilum*, *Lycopodium lucidulum*, *Polypodium vulgare* var. *virginianum*, *Polystichum acrostichoides*, *Dryopteris carthusiana*, *Dryopteris intermedia*, *Dryopteris marginalis*, *Asplenium rhizophyllum*, and *Woodsia obtusa*.

Woodlands

Under this heading are discussed moist, shaded woodlands and open, usually dryish, woodlands. Species listed under the shaded sandstone ferns may be found occasionally in the moist woodlands where they are not associated with rocky outcrops. These species generally are excluded from the discussion in the following paragraph.

MOIST WOODLANDS The deep ravine, generally underlain by sandstone, provides the habitat for the greatest number of fern and fern-ally species in Illinois. Lush canyons in both the northern and southern counties are ideal areas for the lover and student of ferns. Most of the moist woodland ferns are wide-ranging in Illinois and occur in both northern and southern counties. Common species are *Botrychium virginianum*, *Adiantum pedatum*, *Polystichum acrostichoides*, *Onoclea sensibilis*, *Athyrium pycnocarpon*, *Athyrium thelypterioides*, *Athyrium filix-femina* var. *rubellum*, *Asplenium platyneuron*, and *Cystopteris fragilis* var. *protrusa*. Wide-ranging species of moist woodlands which are found occasionally are *Selaginella apoda*, *Ophioglossum vulgatum*, the three species of *Osmunda*, *Thelypteris hexagonoptera*, and *Cystopteris fragilis* var. *fragilis*. Rare species of moist woodlands which are distributed in both northern and southern counties are *Thelypteris phegopteris* and *Thelypteris noveboracensis*.

Only two ferns of moist woodlands are restricted to southern Illinois, and each has been found only a single time. These are *Dryopteris* × *clintoniana* and *Athyrium filix-femina* var. *asplenioides*. On the other hand, four ferns or fern-allies are restricted to northern moist woodlands—*Equisetum scirpoides, Matteuccia struthiopteris, Gymnocarpium dryopteris,* and *Dryopteris goldiana.*

OPEN WOODLANDS The open woodlands are generally dominated by species of oaks and hickories. They usually become rather dry during the summer months. Species regularly found in open woodlands are *Botrychium virginianum, Botrychium dissectum* var. *dissectum, Botrychium dissectum* var. *obliquum, Botrychium biternatum* (from a single southern county), *Botrychium multifidum* ssp. *silaifolium* (from northern Illinois only), *Pteridium aquilinum,* and *Asplenium platyneuron.*

Bogs, Shores, and Standing Water

Most species occupying the aquatic or subaquatic habitats are fern-allies, rather than ferns. The genus *Equisetum* is the most conspicuous group of aquatic fern-allies. Numerous rivers and many lakes and ponds, both natural and artificial, occur in Illinois, providing much area for aquatic species. A few unique, rapidly disappearing sphagnum bogs remain in the northeastern corner of the state.

STANDING WATER Species found in standing water are few indeed. Five native species and one adventive species (*Marsilea quadrifolia*) may be termed true aquatics. The native species are *Equisetum fluviatile, Isoetes melanopoda, Isoetes engelmannii* (from a single station), *Azolla mexicana,* and *Azolla caroliniana* (from a single station).

SHORELINE AND RIVER BANKS Several species of *Equisetum* are the chief pteridophytic inhabitants of shores and banks. These species are tolerant of periodic flooding, but live most of their lives out of water. These species are *Equisetum variegatum, E.* × *nelsonii, E. arvense, E. hyemale* var. *affine, E. laevigatum,* and *E.* × *ferrissii.*

BOGS Species of the northern bogs sometimes occur in great abundance. Two common species are *Osmunda cinnamomea* and *Thelypteris palustris* var. *pubescens.* Species collected rarely from these bogs are *Lycopodium inundatum* and *Woodwardia virginica.*

Fields and Railroad Right-of-Ways

Plants which occupy fields and railroad right-of-ways often are adventive species in an area. The few pteridophytes occurring in such situations are native members of our flora, however. Common in fields, particularly along roads, is *Pteridium aquilinum*. In the gravel embankments of railroad right-of-ways regularly may be found *Equisetum arvense* and *E. hyemale* var. *affine*.

HOW TO IDENTIFY A FERN OR FERN-ALLY

Botanists use written devices known as keys to enable them to identify plants. The keys in this book have been written essentially in nontechnical language in the hope that more people will find them useful. Where technical terms are introduced in the key, they are accompanied by illustrations.

There are two general keys to enable the user to identify the genus of the fern or fern-ally in question. One of these requires that the specimen has sporangia present; the other is designed for use with sterile (vegetative) specimens. Since sporangial structures generally are much less variable than vegetative structures, the key based on sporangia is somewhat more reliable. It begins first with a key to orders, followed by a key to families, followed by a key to genera. Once the genus is ascertained by using one of the general keys, the reader should turn to that genus and use the key provided to the species of that genus if more than one species occurs in Illinois. Of course, if the genus is recognized at sight, then the general keys should be by-passed. The keys in this work are dichotomous—pairs of contrasting statements. Always begin by reading both members of the first pair of characters. By choosing that statement which best fits the specimen to be identified, the reader will be guided to the next proper pair of statements. Eventually, a name will be derived.

Key to the ORDERS of Ferns and Fern-Allies

BASED PARTLY ON FERTILE SPECIMENS

---

1. Leaves grass-like; corm at base of leaves bilobed (**Fig. 21**) ---- -- **Isoetales**, p. 37
1. Leaves not grass-like; corm absent.
   2. Stem conspicuously jointed and with longitudinal ridges (**Fig. 22**) ------------------------------ **Equisetales**, p. 42
   2. Stem not conspicuously jointed and without longitudinal ridges.
      3. Plants true aquatics, either rooted or floating-------- ------------------------------------ **Salviniales**, p. 170
      3. Plants terrestrial although occasional in boggy situations.
         4. Leaves less than 2 cm long, often scale-like.
            5. Creeping plants with erect branches; leaves without a ligule------------------------ **Lycopodiales**, p. 23
            5. Creeping or tufted plants without erect branches; leaves with a ligule near the axil---- **Selaginellales**, p. 35
         4. Leaves more than 2 cm long.
            6. Sporangia borne on a branch arising from base of sterile portion of the blade (**Figs. 23 and 24**)---- ------------------------------ **Ophioglossales**, p. 62
            6. Sporangia borne on the back of or on the margin of green leaves, or on wholly fertile leaves or pinnae (at the apex or center of the leaf) ---------- **Filicales**, p. 75

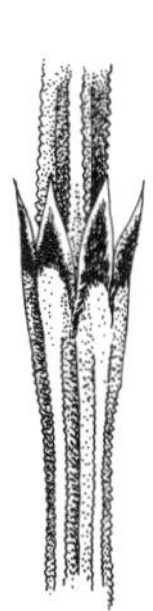

(Left to right) *21. Isoetes. 22. Equisetum. 23. Ophioglossum. 24. Botrychium.*

Key to the GENERA of Ferns and Fern-Allies

BASED ON STERILE SPECIMENS

---

1. Stem longitudinally striate throughout, conspicuously jointed, with sheaths at each joint, the sheaths bearing short teeth (Fig. 22) ________________ **Equisetum,** p. 42
1. Stem neither longitudinally striate nor jointed, without tooth-bearing sheaths.
   2. Leaves very slender, grass-like, needle-like, or scale-like.
      3. Leaves long, grass-like; plant with bilobed corm at base (Fig. 21) ________________ **Isoetes,** p. 37
      3. Leaves shorter, scale-like or soft needle-like; basal corm absent.
         4. Leaves with a small, membranous structure (ligule) near axil ________________ **Selaginella,** p. 35
         4. Leaves lacking a ligule ________________ **Lycopodium,** p. 23
   2. Leaves broader, usually divided, if simple and undivided, then at least 4 mm broad.
      5. True aquatic ferns, rooted or floating in water.
         6. Plants rooted; leaves with 4 pinnae (like a 4-leaf clover) **(Fig. 25)** ________________ **Marsilea,** p. 170

(Left to right)
25. *Marsilea.* 26. *Azolla.* 27. *Asplenium.* 28. *Ophioglossum.*

6. Plants floating; leaves bilobed (**Fig. 26**)____**Azolla**, p. 171

5. Terrestrial ferns, at least under normal environmental conditions.

   7. Leaves simple and entire.

      8. Leaves heart-shaped at base, tapering to a long tip; midvein conspicuous (**Fig. 27**)______**Asplenium**, p. 144

      8. Leaves tapering to base, obtuse or short-acute at tip; midvein absent or inconspicuous (**Fig. 28**)______ ______________________________**Ophioglossum**, p. 70

   7. Leaves simple and pinnatifid to compound.

      9. Leaves pinnatifid (if only the lowest pair or pairs of pinnae are divided to the rachis, see second number 9).

         10. Veins of leaf united to form a network (**Fig. 29**) ______________________________**Onoclea**, p. 107

         10. Veins of leaf free.

            11. Blades translucent, one cell layer thick, glabrous (**Fig. 30**)______________**Trichomanes**, p. 75

            11. Blades more opaque, several cell layers thick, glabrous or pubescent.

               12. Blades deciduous, thin, triangular, the lowest (and largest) pinnae projecting backward (**Figs. 31 and 32**)___**Thelypteris**, p. 111

               12. Blades evergreen, subcoriaceous, elongated, the lowest pinnules not projecting downward.

(Left to right)
29. *Onoclea*. 30. *Trichomanes*. 31. *Thelypteris*. 32. *Thelypteris*.

13. Sinuses between pinnules extending nearly to the winged rachis (**Fig. 33**) ------ ---------------------- **Polypodium,** p. 101

13. Sinuses between pinnules very shallow (**Fig. 34**) ---------------- **Asplenium,** p. 144

9. Leaves compound, with at least the lowest pair of pinnae divided all the way to the rachis.

14. Petiole forked at tip (**Fig. 35**); ultimate segments toothed along upper margin, entire along lower margin (Fig. 4) ------------- **Adiantum,** p. 88

(Left to right)

33. *Polypodium*

34. *Asplenium.*

35. *Adiantum.*

14. Petiole not forked at tip; ultimate segments not with the above pattern on the margins.

15. Blades once-pinnate.

16. Pinnae with an auricle on one margin.

17. Margin of pinnae spiny-toothed; petioles densely scaly (**Fig. 36**) -------- ---------------------- **Polystichum,** p. 104

17. Margin of pinnae not spiny-toothed; petioles not densely scaly (**Fig. 37**) ---- ----------------------- **Asplenium,** p. 144

16. Pinnae without auricles on the margin.

18. Pinnae entire, several times longer than broad (Fig. 16) ------------ **Athyrium,** p. 137

18. Pinnae toothed or lobed, at most only twice as long as broad, usually shorter (**Fig. 38**) ---------------- **Asplenium,** p. 144

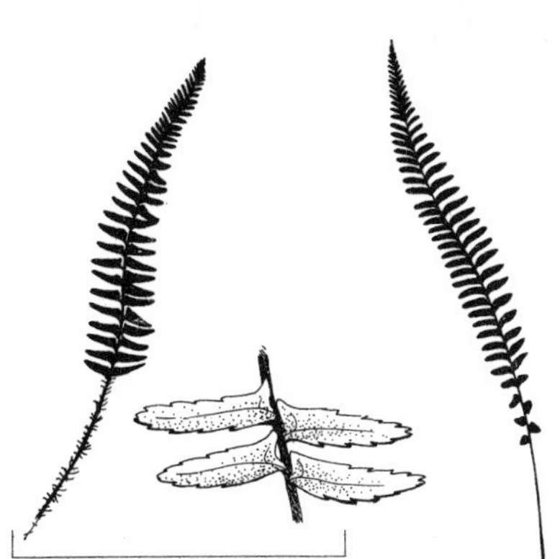
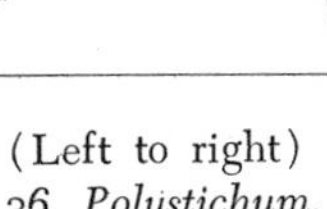

(Left to right)
36. *Polystichum.* 37. *Asplenium.* 38. *Asplenium.* 39. *Botrychium.*

15. Blades pinnate-pinnatifid, bipinnate, bipinnate-pinnatifid, or tripinnate.
 19. Rachis smooth, without scales or hairs.
  20. Blades ternate, appearing broadly triangular.
   21. Ultimate pinnules deeply cut into linear divisions (**Fig. 39**) ---------------------------- **Botrychium**, p. 62
   21. Ultimate pinnules round-lobed or unlobed.
    22. Ultimate pinnules toothed (**Fig. 40**) ------------------- **Botrychium**, p. 62
    22. Ultimate pinnules entire.
     23. Blades delicate, less than 15 cm long (**Fig. 41**) ----------------------- **Gymnocarpium**, p. 108

(Left to right)

40. *Botrychium.*

41. *Gymnocarpium.*

42. *Pteridium.*

23. Blades more firm, always well over 15 cm long (**Fig. 42**)____________________________**Pteridium**, p. 88

20. Blades elongated, not ternate.
    24. Veins of blade forming a single row of closed areoles on either side of midvein (**Fig. 43**)____________________________________**Woodwardia**, p. 134
    24. All veins free.
        25. Lower half of rachis brown, the upper half green____________________________**Asplenium**, p. 144
        25. Rachis all one color, or at least not as above.
            26. Lower pairs of pinnae gradually reduced to minute pinnae (**Fig. 44**)_____________**Matteuccia**, p. 107
            26. Lower pairs of pinnae not much smaller than those above.
                27. Rachis purple-brown____________________________**Pellaea**, p. 94
                27. Rachis brown or green.
                    28. Blades pinnate-pinnatifid or bipinnate.
                        29. Pinnae tapering to a long-pointed apex (**Fig. 45**)____________**Athyrium**, p. 137
                        29. Pinnae not tapering to a long-pointed apex.

43. *Woodwardia.*

44. *Matteuccia.* 45. *Athyrium.* 46. *Osmunda.*

47. *Cryptogramma.*

48. *Osmunda.*

49. *Dryopteris.*

30. Blades glabrous.
    31. Ultimate pinnules finely toothed (**Fig. 46**)______**Osmunda,** p. 76
    31. Ultimate pinnules not finely toothed.
        32. Blades delicate; plants less than 25 cm tall, confined to limestone cliffs (**Fig. 47**)_______________**Cryptogramma,** p. 92
        32. Blades more firm; plants at least 50 cm tall, usually taller, in woods or lowlands, or on sandstone.
            33. Blades pinnate-pinnatifid (**Fig. 48**) __________ **Osmunda,**______p. 76
            33. Blades bipinnate (**Fig. 49**)__________**Dryopteris,** p. 119
30. Blades hairy.
    34. Fronds up to 75 cm long___**Thelypteris,** p. 111

50. *Dryopteris.*

51. *Athyrium.*

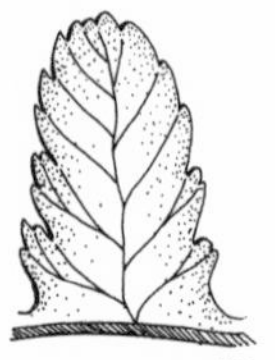
52. *Cystopteris.*

53. *Woodsia.*

34. Fronds generally always well over 75 cm long_____**Osmunda,** p. 76

28. Blades bipinnate-pinnatifid to tripinnate.

35. Ultimate pinnules with bristly-tipped teeth (**Fig. 50**).

36. Veins running to the tip of the teeth (**Fig. 51**)_______**Athyrium,** p. 137

36. Veins stopping at the base of the teeth (Fig. 50)_______________**Dryopteris,** p. 119

35. Ultimate pinnules lobed or round-toothed (**Fig. 52**).

37. Veins of blade extending to the sinus between the teeth (Fig. 52)________________**Cystopteris,** p. 160

37. Veins of blade extending to tip of teeth of lobes (Fig. 51).

38. Blade subcoriaceous, evergreen_______________**Dryopteris,** p. 119

38. Blades thinner, deciduous.

39. Plants generally less than 30 cm tall _____**Cystopteris,** p. 160

39. Plants generally more than 30 cm tall, usually at least 60 cm tall____________**Athyrium,** p. 137

19. Rachis hairy or scaly or glandular.

40. Rachis both hairy and scaly.

41. Blades triangular, the rachis winged nearly to base (Fig. 31)_____________________________**Thelypteris,** p. 111

41. Blades elongated, the rachis winged only near apex, or unwinged (**Fig. 53**) _____________________**Woodsia,** p. 159

(Left to right)

54. *Dennstaedtia.*

55. *Woodsia.*

56. *Cheilanthes.*

40. Rachis hairy or scaly, but not both.
  42. Rachis glandular.
    43. Petiole and rachis hairy (**Fig. 54**). __________________**Dennstaedtia,** p. 86
    43. Petiole and rachis scaly (**Fig. 55**) ______________________**Woodsia,** p. 159
  42. Rachis without glands.
    44. Rachis hairy.
      45. Rachis purple-brown.
        46. Blades densely hairy (**Fig. 56**) _________________**Cheilanthes,** p. 97
        46. Blades glabrous or nearly so (**Fig. 57**)___________**Pellaea,** p. 94
      45. Rachis not purple-brown.
        47. Blades ternate, appearing triangular.
          48. Ultimate pinnules glabrous _____________**Botrychium,** p. 62
          48. Ultimate pinnules hairy ______________**Pteridium,** p. 88
        47. Blades elongated, not ternate.
          49. Pinnae with a tuft of brown wool in their axils (**Fig. 58**) ______________**Osmunda,** p. 76
          49. Pinnae without a tuft of brown wool in their axils.
            50. Lower pair of pinnae gradually reduced in size (Fig. 44) ___**Matteuccia,** p. 107

57. *Pellaea.*

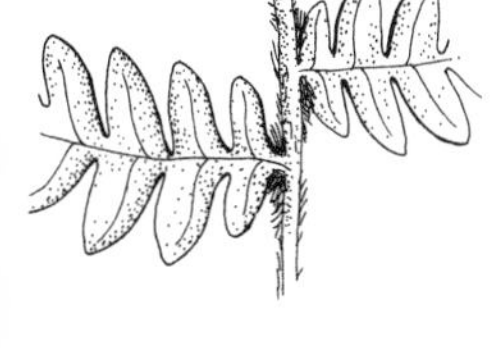

58. *Osmunda.*

59. *Dryopteris.*

50. Lower pinnae not much smaller than those above (**Fig. 59**)____**Dryopteris,** p. 119

44. Rachis scaly.
  51. Blade pinnate-pinnatifid or bipinnate.
    52. Blade subcoriaceous, evergreen _______________**Dryopteris,** p. 119
    52. Blade thinner, deciduous.
      53. Blade distinctly triangular; all but the lowest pair of pinnae connected by a winged rachis (Fig. 31) _____________**Thelypteris,** p. 111
      53. Blade not triangular; most of the pinnae stalked (Fig. 18)______**Athyrium,** p. 137
  51. Blade bipinnate-pinnatifid to tripinnate.
    54. Ultimate pinnules with bristly-tipped teeth (Fig. 50).
      55. Veins running to the tip of the teeth (Fig. 51)________ ______________**Athyrium,** p. 137
      55. Veins stopping at the base of the teeth (Fig. 50)______ _____________**Dryopteris,** p. 119
    54. Ultimate pinnules lobed or round-toothed (Fig. 51)____ _______________**Athyrium,** p. 137

Descriptions and Illustrations

Order Lycopodiales

Only the following family comprises this order.

LYCOPODIACEÆ – CLUBMOSS FAMILY

Only the following genus occurs in Illinois.

1. *Lycopodium* [Dill.] L. – Clubmoss

Creeping and sometimes erect plants with simple or much-branched stems usually densely clothed with small, narrowed, evergreen leaves; leaves 1-nerved, 4- to 10-ranked; leaf gaps absent; sporangia axillary or aggregated in terminal cones, 1-locular; spores uniform, 4-sided, papillate or reticulate (in the Illinois species).

For a revision of North American species, see Lloyd and Underwood (1900) and, for the Complanata group, see Wilce (1965).

KEY TO THE SPECIES OF Lycopodium IN ILLINOIS

1. Sporangia axillary along the stem, the leaves subtending the sporangia appearing no different from ordinary leaves; gemmae frequently present in the axils of upper leaves.
    2. Leaves linear or lanceolate, broadest near base, entire; stomates present on both surfaces (with 10× magnification); spores usually over 35 μ in diameter_______________1. *L. porophilum*
    2. Leaves oblanceolate, broadest near middle, serrulate or rarely entire; stomates present only on lower surface (with 10× magnification); spores usually less than 35 μ in diameter____2. *L. lucidulum*
1. Sporangia aggregated in terminal cones, the leaves subtending the sporangia much modified from ordinary leaves (except in *L. inundatum*); gemmae absent.
    3. Leaves subtending sporangia similar to ordinary leaves, green; sterile stems prostrate_____________________3. *L. inundatum*

3. Leaves subtending sporangia much shorter than ordinary leaves, yellow; sterile stems erect or ascending.
    4. Cones solitary, 5–7 mm thick; leaves 6- to 8-ranked -- 4. *L. dendroideum*
    4. Cones 2–4, 3–6 mm thick; leaves 4-ranked.
        5. Peduncles 3–10 cm long; leaves with appressed tips, 1–2 mm apart; sporophylls without scarious margins, ovate ---------------------------------- 5. *L. flabelliforme*
        5. Peduncles 12–18 cm long; leaves with spreading tips, about 4 mm apart; sporophylls with scarious margins, orbicular ---------------------------- 6. *L. × habereri*

1. Lycopodium porophilum Lloyd & Underw. in Bull. Torrey Club 27:150. 1900. *Fig. 60.*

Lycopodium selago var. *porophilum* (Lloyd & Underw.) Clute, in Fern Bull. 11:47. 1903.

Tufted plant with weakly or strongly ascending stems, unbranched or rarely forking; leaves linear-lanceolate, attenuate at apex, broadest near base, entire, spreading, 4–8 mm long, with stomates present on both surfaces; gemmae frequently present in axils of upper leaves; sporangia axillary, subtended by leafy sporophylls; spores papillate, 35–50 μ in diameter.

COMMON NAME: Cliff Clubmoss.

HABITAT: On moist, shaded sandstone cliffs.

RANGE: Ontario to Minnesota, south to Missouri, Alabama, and South Carolina.

ILLINOIS DISTRIBUTION: Rare; known only from three northern counties and three extreme southeastern counties.

Although Wilson (1932) has argued that *L. porophilum* is equivalent to *L. selago* var. *patens,* Wherry (1961) has shown that this latter taxon is distinct and ranges to the north of Illinois. Wilson's smooth-margined var. *occidentale,* based on Clute's f. *occidentale,* properly belongs to *L. selago* and not to *L. lucidulum.*

2. Lycopodium lucidulum Michx. Fl. Bor. Am. 2:284. 1803.

Tufted plant with more or less ascending stems, usually unbranched (except at base); leaves oblanceolate, attenuate at apex, broadest near middle, serrulate in the upper half or entire, somewhat spreading, 7–14 mm long, with stomates confined to

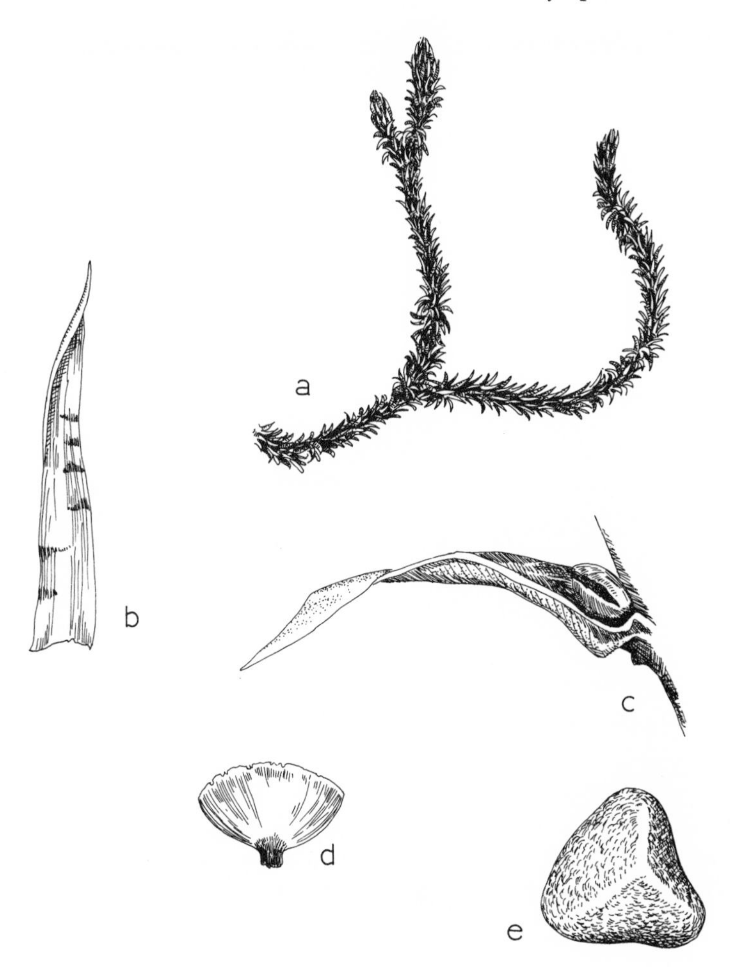

60. *Lycopodium porophilum* (Cliff Clubmoss). *a.* Habit, X½; *b.* Leaf, X7½; *c.* Leaf with axillary sporangium, X10; *d.* Spore, X500; *e.* Spore, X500.

the lower surface; gemmae frequently present in axils of upper leaves; sporangia axillary, subtended by leafy sporophylls; spores papillate, 20–32 μ in diameter.

Two varieties occur in Illinois, distinguished by the following key:

a. Leaves serrulate in the upper half__2a. *L. lucidulum* var. *lucidulum*

a. Leaves entire____________________2b. *L. lucidulum* var. *tryonii*

2a. Lycopodium lucidulum Michx. var. **lucidulum** *Fig. 61.*
Leaves serrulate in the upper half.

COMMON NAME: Shining Clubmoss.
HABITAT: Shaded sandstone cliffs and ravines.
RANGE: Newfoundland to Ontario, south to Missouri and South Carolina.
ILLINOIS DISTRIBUTION: Rare; scattered throughout the state, except the western counties.

2b. Lycopodium lucidulum Michx. var. **tryonii** Mohlenbrock, var. nov. *Fig. 61e.*
Folia integra.

TYPE: *Hatcher s.n.*, from Little Grand Canyon, Jackson County, Illinois.
COMMON NAME: Shining Clubmoss.
HABITAT: Shaded sandstone cliffs and ravines.
RANGE: Same as var. *lucidulum.*
ILLINOIS DISTRIBUTION: Very rare; known only from Jackson County (in S.I.U. herbarium).
Although Wilson (following Clute) created *L. lucidulum* var. *occidentale* as an entire-leaved taxon, Clute's epithet belongs to *L. selago.*
The epithet commemorates Rolla Milton Tryon, ardent student of ferns, who first suggested to the writer several years ago that the entire-margined variety of *L. lucidulum* was in need of an epithet.

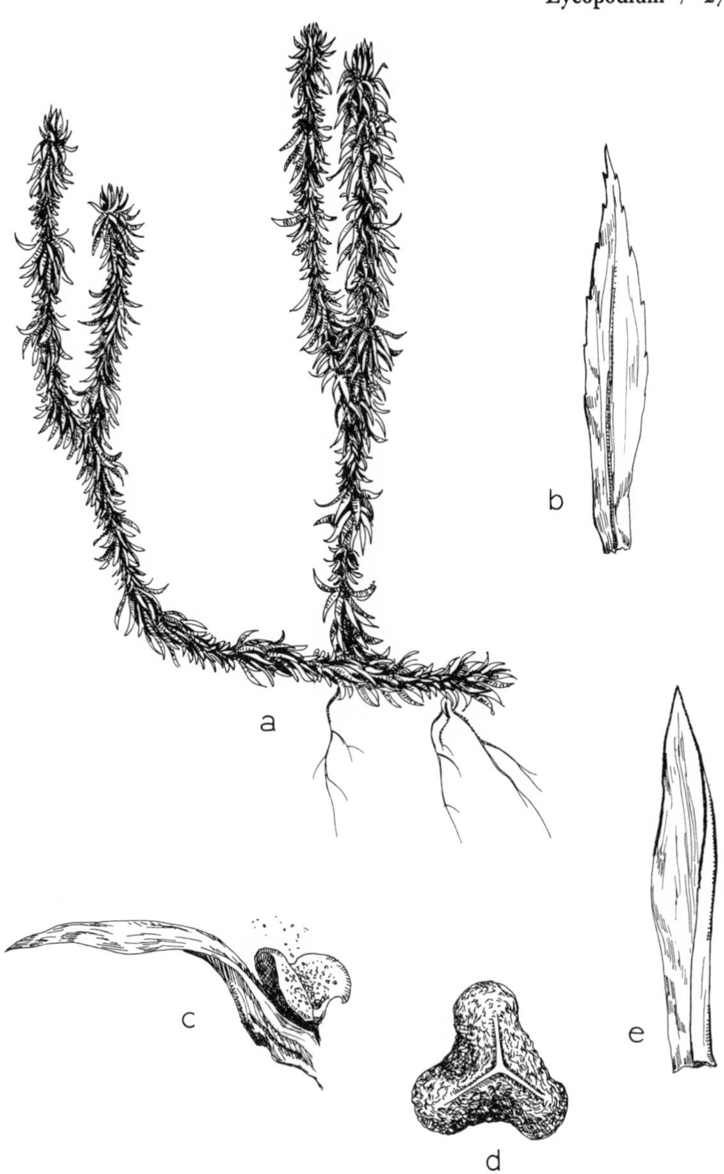

61. *Lycopodium lucidulum* (Shining Clubmoss). var. *lucidulum*—*a.* Habit, X½; *b.* Leaf, X5; *c.* Leaf with axillary sporangium, X5; *d.* Spore, X660; var. *tryonii*—*e.* Leaf, X5.

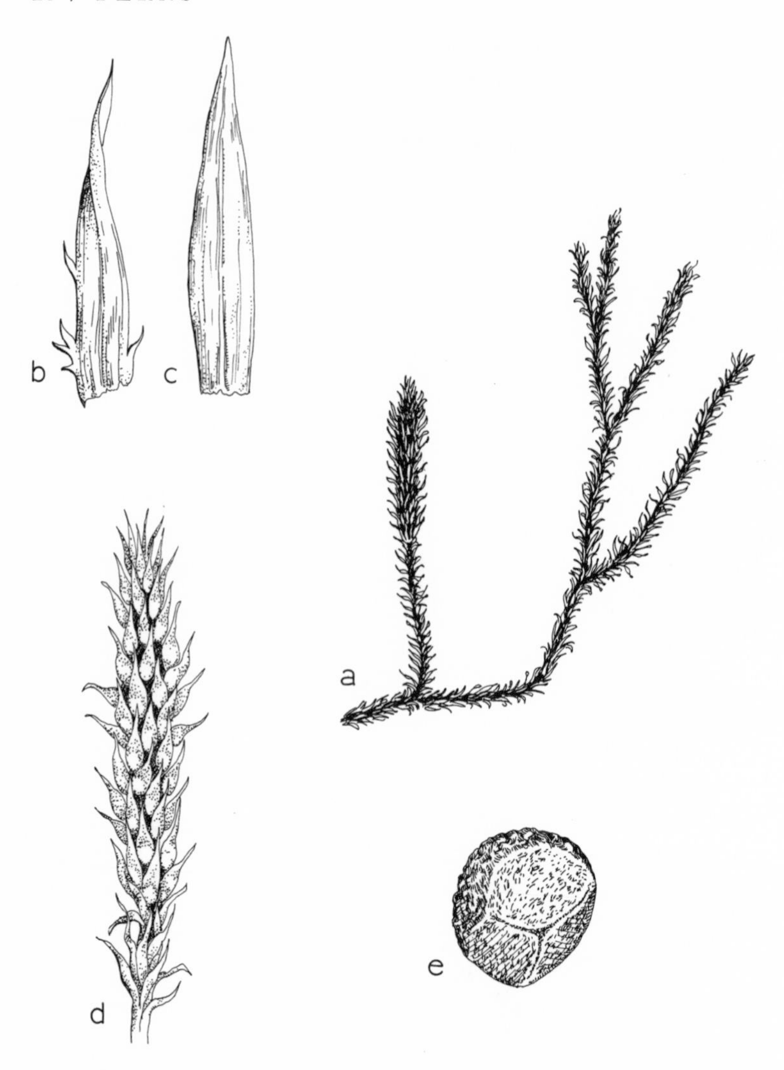

62. *Lycopodium inundatum* (Bog Clubmoss). *a.* Habit, X½, *b.*, *c.* Leaves, X7½; *d.* Cone, X½; *e.* Spore, X330.

3. Lycopodium inundatum L. Sp. Pl. 1102. 1753. *Fig. 62.*

Extensively creeping plant with forked, prostrate stems; leaves linear-lanceolate, attenuate at apex, entire or spinulose, somewhat spreading, 8- to 10-ranked, 5–8 mm long; gemmae absent; sporangia borne in a terminal cone; cone 1–4 cm long, 0.5–1.0 cm thick; sporophylls green, leafy, lanceolate, entire or serrulate near base; spores papillate at apex, reticulate at base, 45–55 μ in diameter; n = 78.

COMMON NAME: Bog Clubmoss.

HABITAT: Bogs.

RANGE: Newfoundland to Alaska, south to Oregon, Illinois, and Virginia.

ILLINOIS DISTRIBUTION: Very rare; known only from Cook County. The most recent collection is from near Thornton, September 14, 1947, by *J. A. Steyermark 65010*.

This species can be considered transitional between *L. porophilum* and *L. lucidulum* on the one hand and *L. dendroideum* and *L. flabelliforme* on the other. It has leafy sporophylls like the first two, but the sporophylls are arranged in a terminal cone like the last two.

4. **Lycopodium dendroideum** Michx. Fl. Bor. Am. 2:282. 1803. *Fig. 63.*

Lycopodium obscurum L. var. *dendroideum* (Michx.) D. C. Eaton ex A. Gray, Man. 696. 1890.

Creeping plant with ascending, branched, sterile stems; leaves 6- to 8-ranked, linear-lanceolate, attenuate at the apex, entire, 4–6 mm long, the lower appressed, the upper spreading; gemmae absent; sporangia borne in a terminal cone; cone solitary, sessile, 1.5–5.5 cm long, 5–7 mm thick; sporophylls yellow, much reduced, acuminate at apex, cordate; spores faintly reticulate, 25–35 μ in diameter.

COMMON NAME: Ground Pine.

HABITAT: Woodlands.

RANGE: Labrador to Alaska, south to Montana, Illinois, and Georgia.

ILLINOIS DISTRIBUTION: Very rare; known only from Ogle County (moist, sandy detritus of St. Peter sandstone, Castle Rock, south of Oregon, *Winterringer & Hooks 17187*) in northern Illinois. This taxon is considered by some to be merely a variety of *L. obscurum*. There is justification for recognizing it as a distinct species because of its narrower, incurved leaves and its more slender cones.

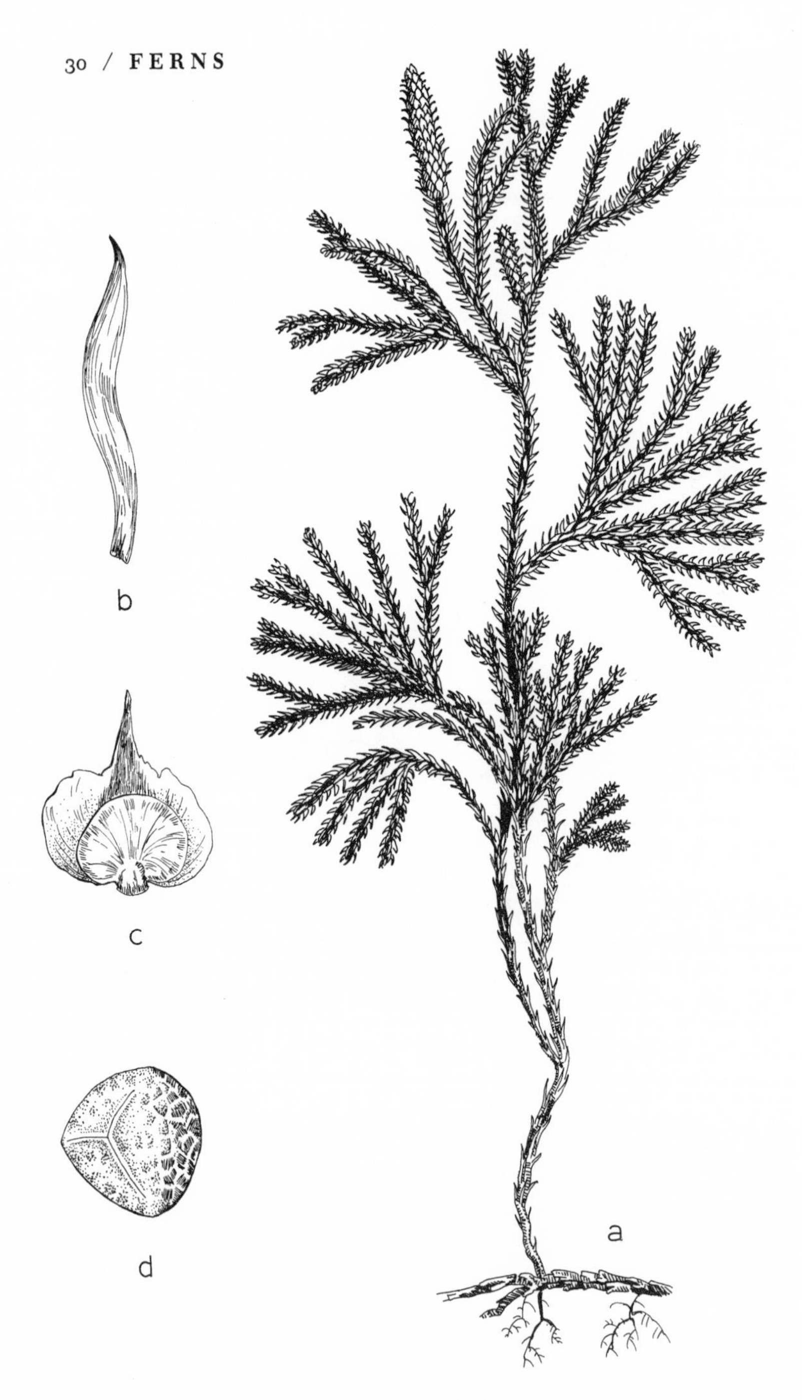

63. *Lycopodium dendroideum* (Ground Pine). *a.* Habit, X½; *b.* Leaf, X7½; *c.* Sporangium and leaf, X6; *d.* Spore, X500.

5. Lycopodium flabelliforme (Fern.) Blanch. in Rhodora 13:168. 1911. *Fig. 64.*

Lycopodium complanatum L. var. *flabelliforme* Fern. in Rhodora 3:280. 1901.

Creeping plant with ascending, dichotomously branched, sterile stems; leaves 4-ranked, small, often subulate, entire; gemmae absent; sporangia borne in terminal cones; cones 3–4, pedunculate, 2–5 cm long, 3–6 mm thick, the peduncles 3–10 cm long; sporophylls yellow, much reduced, broad at base; spores reticulate on the outer faces, 28–36 μ in diameter.

COMMON NAME: Ground Pine.

HABITAT: Exposed sandstone ledges and low areas.

RANGE: Newfoundland to Ontario, south to Iowa, Illinois, and South Carolina.

ILLINOIS DISTRIBUTION: Two native stations (Pope: sandstone ledge, Lusk Creek, *J. Swayne 5780;* Ogle: cold, damp, sandy detritus of St. Peter sandstone, Muir Creek, *S. Spongberg 63/166*) and three adventive stations (DuPage: under *Picea abies,* Morton Arboretum, *E. L. Kammerer s.n.;* Cook: Ned Brown Forest Preserve, *V. Schwarz & T. Brodene 2242;* Cook: Dan McMahon Forest Preserve, *Evers 77193*).

Fernald originally described this taxon as a variety of *L. complanatum,* and this view has been followed by many authors. However, the rhizome and the pattern of branching seem to indicate that this taxon is a good species, a view concurred by Wagner and Wilce. Dr. Wagner has checked the identification of the Pope County specimens.

6. Lycopodium × habereri House, in N. Y. State Mus. Bull. 176:36. 1915. *Fig. 64f.*

Stems procumbent, to 50 cm long; leaves 4-ranked, with decurrent adnate bases, awl-shaped, the lateral ones with spinulose spreading tips and about 4 mm apart, the ventral ones almost obsolete, 1 mm long; peduncles 12–18 cm long, with 2–4 cones; scales on peduncle 2–3 mm long, subulate, 1.5 cm apart; sporophylls pale green, orbicular, scarious-margined, 1.5–2.0 mm broad.

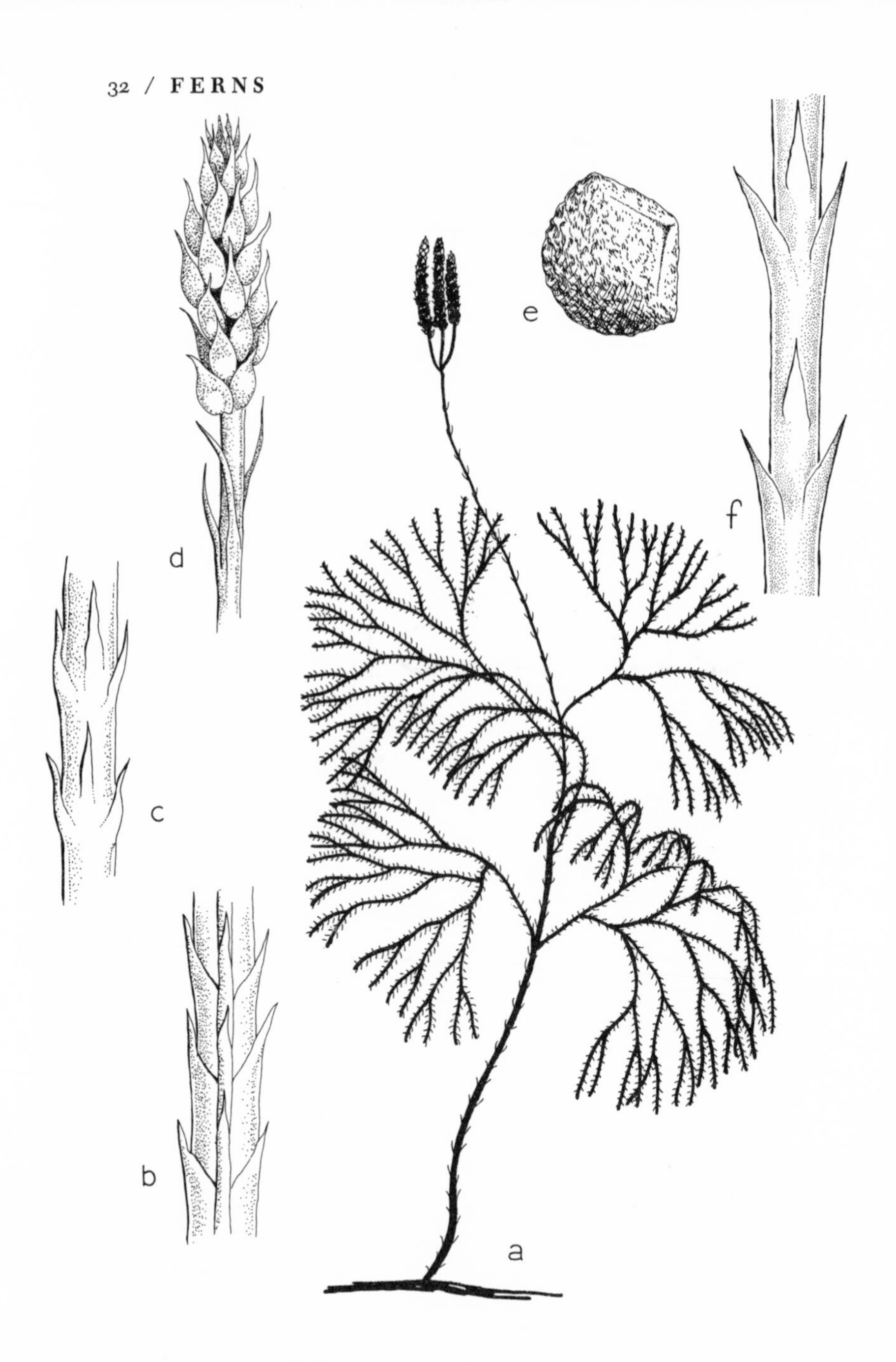

64. *Lycopodium flabelliforme* (Ground Pine). *a.* Habit, X½; *b.*, *c.* Leaves and stems, X7½; *d.* Cone, X2; *e.* Spore, X500; *f. Lycopodium* × *habereri* (Ground Pine), leaves and stems, X7½.

COMMON NAME: Ground Pine.

HABITAT: Shaded cliffs.

RANGE: Newfoundland to Quebec; New York to Illinois.

ILLINOIS DISTRIBUTION: Known only from a single collection from Cook County (Evanston, *Kagey s.n.* in 1878).

This is a reputed hybrid between *L. flabelliforme* and *L. tristachyum*. It may be distinguished from *L. flabelliforme* by its distant leaves with spreading tips and its green, orbicular sporophylls with scarious margins.

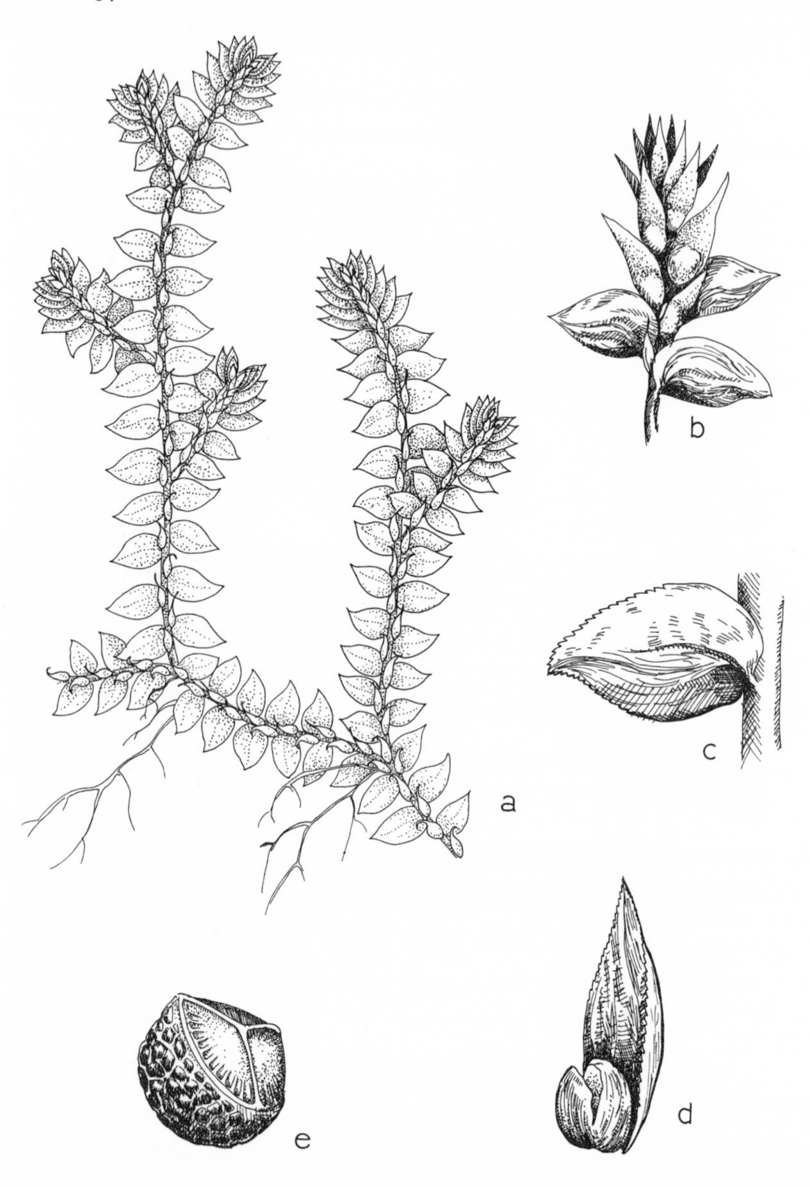

65. *Selaginella apoda* (Small Spikemoss). *a.* Habit, X5; *b.* Cone, X7½; *c.* Leaf, X12½; *d.* Leaf with axillary sporangium, X15; *e.* Spore, X350.

Order Selaginellales

Only the following family comprises this order.

SELAGINELLACEÆ – SPIKEMOSS FAMILY

Only the following genus occurs in Illinois.

1. *Selaginella* BEAUV. – Spikemoss

Small, prostrate plants with branched stems clothed with small, often evergreen, leaves; leaves 4- to 6-ranked; leaf gaps absent; sporangia axillary, solitary, unilocular, of two kinds.

KEY TO THE SPECIES OF **Selaginella** IN ILLINOIS

1. Stems weak, herbaceous; leaves 4-ranked, flaccid, obtuse to acute -- 1. S. *apoda*
1. Stems wiry, evergreen; leaves spirally arranged, stiff, subulate-tipped-- 2. S. *rupestris*

1. Selaginella apoda (L.) Fern. in Rhodora 17:68. 1915. *Fig.* 65.

Lycopodium apodum L. Sp. Pl. 1105. 1753.
Selaginella apus (L.) Spring ex Mart. Fl. Bras. 1 (2):119. 1840.

Small, delicate creeper with much-branched stems; leaves membranous, 4-ranked, the larger ones more or less oblong and acute, spreading, 1–2 mm long, 0.5–1.0 mm broad, ciliolate on the margin, the smaller ones narrower and appressed; cone spike-like, appearing sessile, 10–18 mm long; sporangia 0.7 mm long; sporophylls oblong, similar to the spreading leaves; megaspores reticulate only on the outer face, a little less than 0.5 mm in diameter; microspores orange.

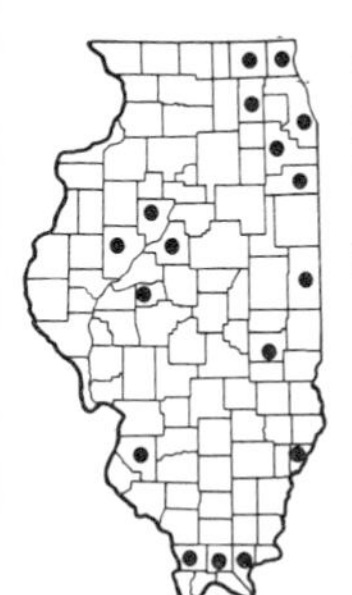

COMMON NAME: Small Spikemoss.
HABITAT: Moist shaded areas, particularly on sandstone boulders.
RANGE: Maine to Ontario, south to Texas and Florida; South America.
ILLINOIS DISTRIBUTION: Local throughout the state.
This small moisture-loving species is frequently confused with species of mosses which occupy similar habitats.

2. Selaginella rupestris (L.) Spring, in Flora 21:149. 1838. *Fig. 66.*

Lycopodium rupestre L. Sp. Pl. 1101. 1753.

Small, wiry creeper with much-branched stems; leaves rigid, spirally arranged, subulate, acuminate, bristly-tipped, appressed, 1–2 mm long, less than 0.5 mm broad, ciliolate on the margin; cone spike-like, more or less 4-angled, 10–15 mm long; sporangia 0.5 mm long; sporophylls ovate, denticulate, bristle-tipped; megaspores reticulate on all faces, 0.5 mm in diameter.

COMMON NAME: Rock Spikemoss.

HABITAT: Dry rocky areas of sandstone.

RANGE: Nova Scotia to Ontario, south to Oklahoma and Georgia.

ILLINOIS DISTRIBUTION: Local; mostly in northern Illinois; also Pope County (exposed sandstone bluff, Burden Falls, and Jackson Hollow) in extreme southeastern Illinois.

For a discussion of this species and its variations, see Tryon (1955).

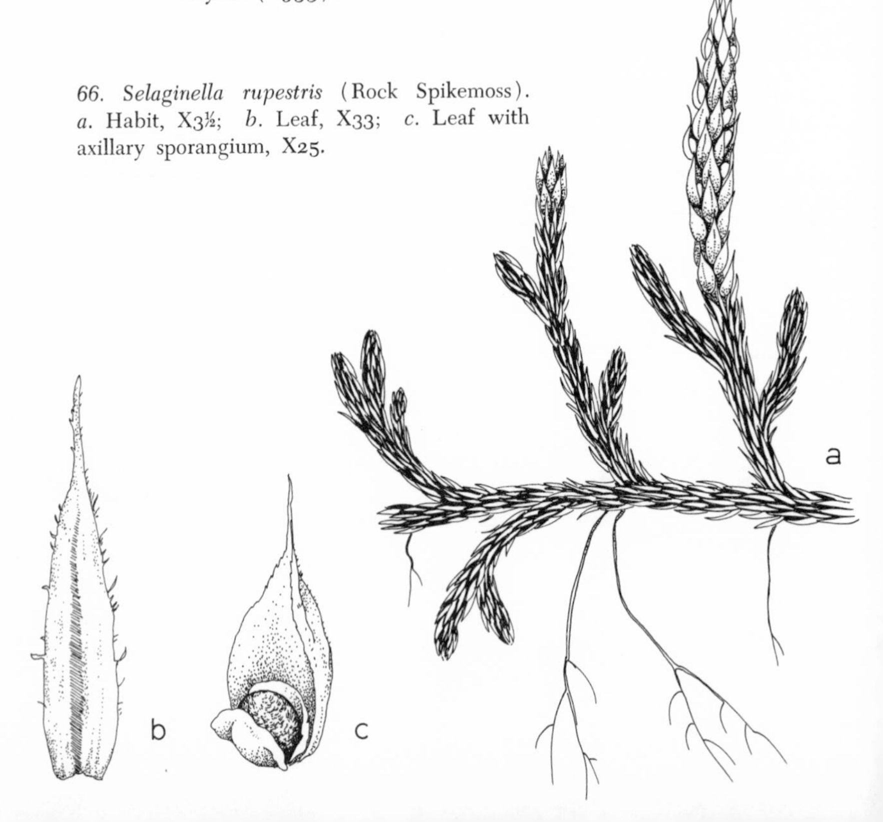

66. *Selaginella rupestris* (Rock Spikemoss). *a.* Habit, X3½; *b.* Leaf, X33; *c.* Leaf with axillary sporangium, X25.

Order Isoetales

Only the following family comprises this order.

ISOETACEÆ – QUILLWORT FAMILY

Only the following genus occurs in Illinois.

1. *Isoetes* L. – Quillwort

Stem a fleshy, lobed corm; leaves grass-like, spirally arranged, broadened at base; leaf gaps absent; ligule present just above sporangium; sporangia embedded in leaf bases, brown-punctate or -striate, partially covered above; spores of two kinds.

The most recent monograph of the genus is by Pfeiffer (1922).

KEY TO THE SPECIES OF Isoetes IN ILLINOIS

1. Sporangium punctate or striate; megaspores tuberculate; microspores spinulose or papillose.
   2. Leaves mostly 15–40 cm long; ligule subulate, triangular; sporangia punctate; megaspores 280–440 μ in diameter; microspores 20–30 μ in diameter, spinulose _________ 1. *I. melanopoda*
   2. Leaves mostly 8–15 cm long; ligule elongate, cordate; sporangia striate; megaspores 480–650 μ in diameter; microspores 27–37 μ in diameter, papillose _________ 2. *I. butleri*
1. Sporangium neither punctate nor striate; megaspores reticulate, with narrow ridges; microspores smooth or minutely roughened _________ 3. *I. engelmannii*

1. Isoetes melanopoda Gay & Dur. in Bull. Soc. Bot. Fr. 11:102. 1864. *Fig. 67.*

Isoetes melanopoda var. *pallida* Engelm. in Trans. St. Louis Acad. Sci. 4:387. 1882.

Isoetes melanopoda f. *pallida* (Engelm.) Fern. in Rhodora 51:103. 1949.

Corm 2-lobed; leaves 15–85, 15–40 cm long, black and shining at base, or occasionally pale (f. *pallida*); ligule subulate, triangular; sporangia oblongoid, 0.5–3.0 cm long, brown-punctate; megaspores 280–440 μ in diameter, low-tuberculate; microspores 20–30 μ in diameter, spinulose.

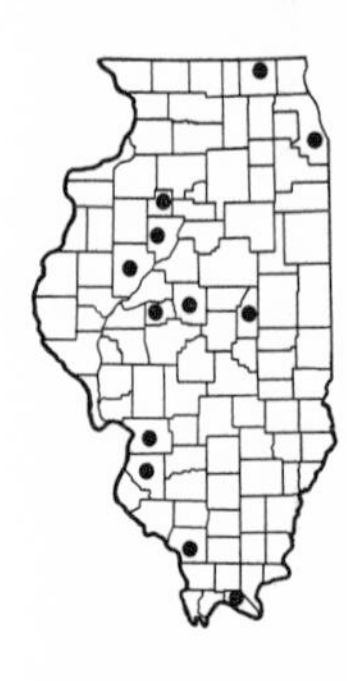

COMMON NAME: Black Quillwort.

HABITAT: Shallow water of ponds and ditches.

RANGE: Illinois and Iowa, south to Texas and Georgia; New Jersey.

ILLINOIS DISTRIBUTION: Local throughout Illinois. The type is from Menard County.

The pale-based form may be found growing with the black ones. *Isoetes melanopoda* differs from *I. butleri* not only in its larger measurements but in the configuration of the megaspores and microspores.

2. **Isoetes butleri** Engelm. in Bot. Gaz. 3:1. 1878. *Fig.* 68.

Corm 2-lobed; leaves 8–30, 8–15 cm long, slender, usually dark at base; ligule elongate, cordate; sporangia oblongoid, 6–7 mm long, brown-striate; megaspores 480–650 μ in diameter, tuberculate; microspores 27–37 μ in diameter, papillose.

COMMON NAME: Butler's Quillwort.

HABITAT: Moist depressions in limestone or sandstone (in Illinois) areas.

RANGE: Illinois to Kansas, south to Oklahoma and Arkansas.

ILLINOIS DISTRIBUTION: Local; confined to the extreme southern counties in the Shawnee Hills, on sandstone.

Some botanists have suggested that this species is probably only a variety of *I. melanopoda,* but there are no experimental data to back this up.

In Missouri, where this species is frequent on Ozark glades, the chief rock-type is limestone.

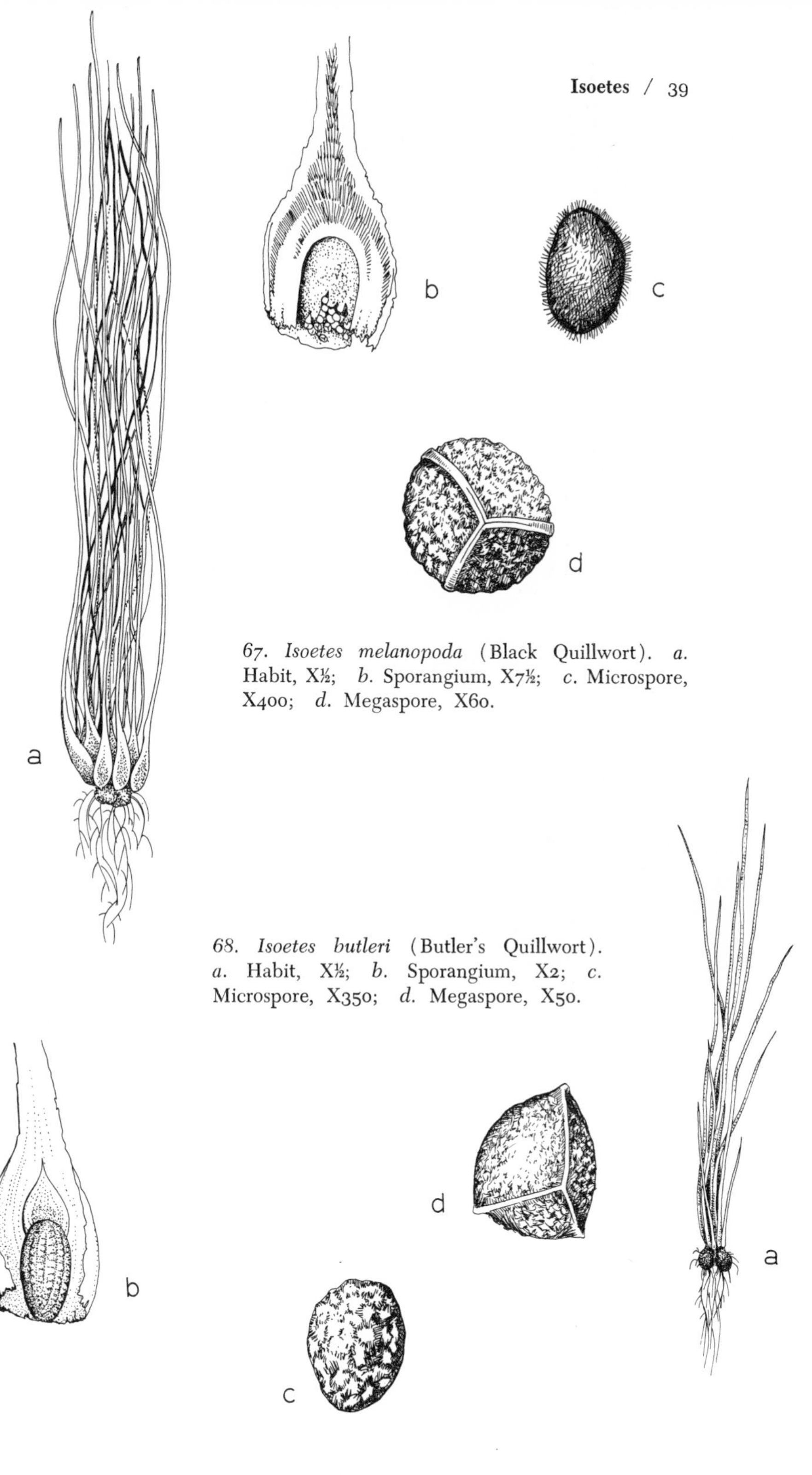

67. *Isoetes melanopoda* (Black Quillwort). *a.* Habit, X½; *b.* Sporangium, X7½; *c.* Microspore, X400; *d.* Megaspore, X60.

68. *Isoetes butleri* (Butler's Quillwort). *a.* Habit, X½; *b.* Sporangium, X2; *c.* Microspore, X350; *d.* Megaspore, X50.

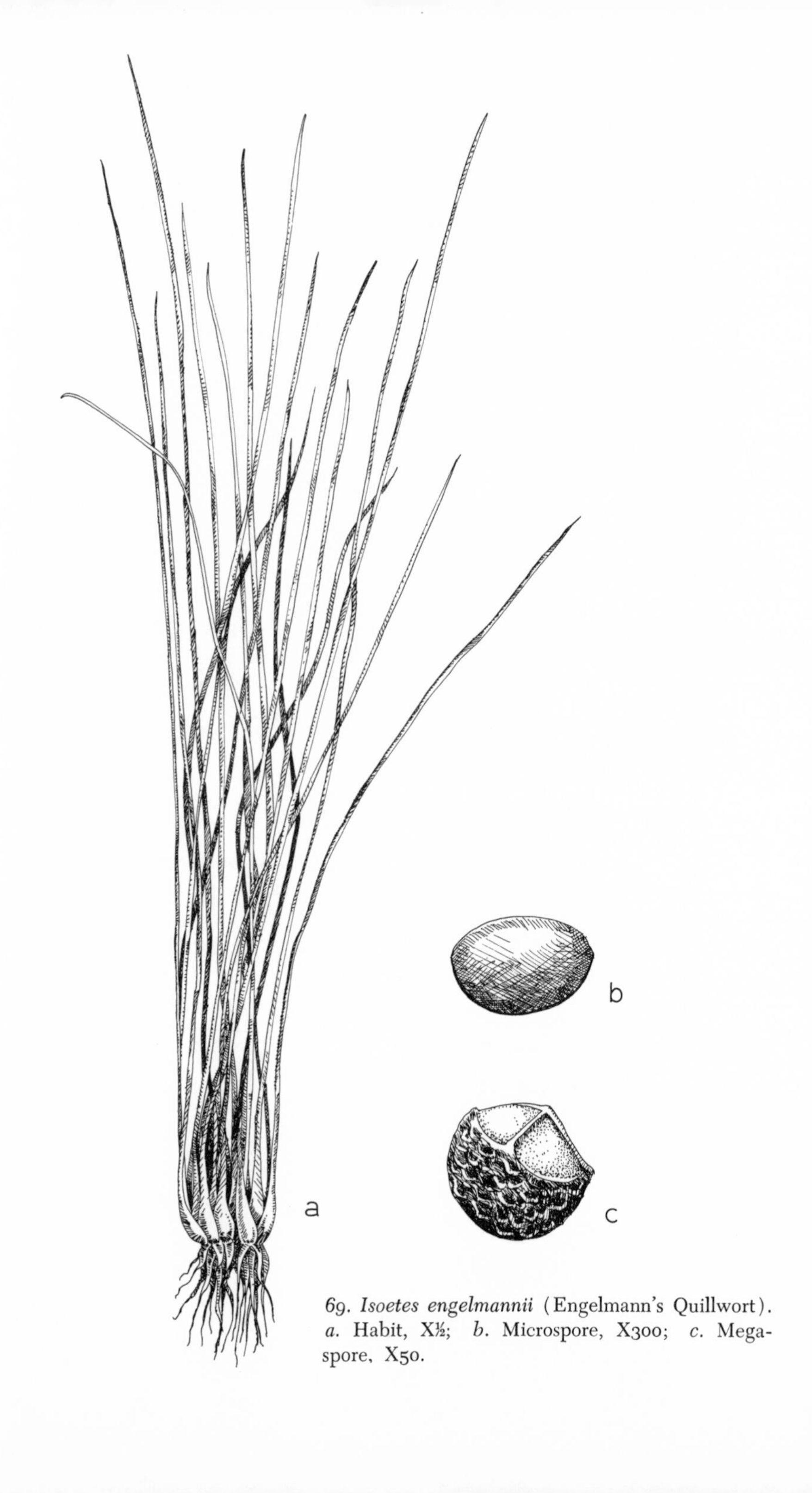

69. *Isoetes engelmannii* (Engelmann's Quillwort). *a*. Habit, X½; *b*. Microspore, X300; *c*. Megaspore, X50.

3. **Isoetes engelmannii** A. Br. in Flora 29:178. 1846. *Fig.* 69.

Corm 2-lobed; leaves 15–60 (–100), 13–50 cm long, slender; ligule triangular; sporangia oblongoid, 6–13 mm long, neither punctate nor striate; megaspores 400–615 μ in diameter, reticulate with narrow ridges; microspores 21–33 μ in diameter, smooth to minutely roughened.

COMMON NAME: Engelmann's Quillwort.

HABITAT: Shallow water of ponds and ditches.

RANGE: New Hampshire to Missouri, south to Alabama and Florida.

ILLINOIS DISTRIBUTION: Very rare and probably now extinct; known only from St. Clair County (ponds, rare, *H. Eggert*).

Sporangium, megaspore, and microspore characters distinguish this species, but particularly the nearly smooth microspores.

Order Equisetales

Only the following family comprises this order.

EQUISETACEÆ – HORSETAIL FAMILY

Only the following genus comprises this family.

1. Equisetum [Tourn.] L. – Horsetail

Rhizome stout, branched; aerial stems erect, hollow, ridged, conspicuously jointed at the nodes; branches and leaves borne in whorls at the nodes, the leaves composing the sheath; sporangia borne in terminal cones; spores alike.

This genus occurs throughout the world and is made up of about twenty-five species. The hollow, jointed stems may be branched or unbranched, and both conditions may occur in the same species. A sheath is present at each joint on the stem and usually bears slender teeth. Between the joints, the stem is ridged and bears 1–2 rows of silica tubercles. The stems commonly contain silica deposits. Cones bearing the sporangia may be borne atop either branched or unbranched stems. The genus is considered to be a relatively primitive one, and fossil representatives of its ancestors commonly are found in Illinois.

The most recent treatment of the species is by Hauke (1962).

KEY TO THE SPECIES OF Equisetum IN ILLINOIS

1. Aerial stems brown, fertile______________________6. *E. arvense*
1. Aerial stems green, fertile or sterile.
    2. Ridges of stem 3–6; teeth of sheaths 3; central cavity of stem absent________________________________1. *E. scirpoides*
    2. Ridges of stem usually more than 6; teeth of at least some sheaths more than 3; central cavity of stem present.
        3. Central cavity up to two-thirds the diameter of the stem.
            4. Stems unbranched above the base.
                5. Ridges of stem angled, with tubercles in two rows; teeth of sheath with a central groove throughout; stems firm, perennial.
                    6. Ridges of stem strongly grooved, 4–10-angled; teeth of sheaths falling off early; uppermost sheath (including deciduous teeth) rarely more than 5 mm long________________________2. *E. variegatum*

6. Ridges of stem slightly grooved, 10–17-angled; teeth of sheaths persistent; uppermost sheath (including teeth) nearly always more than 5 mm long______________________3. *E.* × *trachyodon*

5. Ridges of stem rounded, with tubercles in one row; teeth of sheath with a central groove only at tip; stems flexible, annual__________________4. *E.* × *nelsonii*

4. Stems regularly branched above the base.

7. Sheath of first joint of branches with 5–10 teeth; stem with outer row of cavities about same size as central cavity_____________________________5. *E. palustre*

7. Sheath of first joint of branches with 3–4 teeth; stem with outer row of cavities considerably smaller than central cavity.

8. Fertile stems brown; spores normal____6. *E. arvense*

8. Fertile stems green; spores aborted____7. *E.* × *litorale*

3. Central cavity more than two-thirds the diameter of the stem.

9. Teeth of sheaths early deciduous, basally connate.

10. All sheaths gray__________________8. *E. hyemale*

10. Upper sheaths green, the lower sometimes gray.

11. Stems smooth; central cavity three-fourths the diameter of the stem; cone rounded or apiculate at summit______________________9. *E. laevigatum*

11. Stems slightly rough; central cavity four-fifths the diameter of the stem; cone apiculate at summit ____________________________10. *E.* × *ferrissii*

9. Teeth of sheaths persistent, free or basally connate.

12. Teeth of sheaths free; sheaths brownish or green; stems often branched; cones not apiculate _______________________________________11. *E. fluviatile*

12. Teeth of sheaths basally connate; sheaths gray; stems usually unbranched; cones apiculate____8. *E. hyemale*

1. Equisetum scirpoides Michx. Fl. Bor. Am. 2:281. 1803. *Fig. 70.*

Stems evergreen, very slender, usually unbranched, to 15 cm tall, curled, with 3–6 smooth ridges, without a central cavity but with 3 cavities between center of stem and outer wall; sheaths green below, dark above, with three teeth, the tip of each tooth usually deciduous; cones 2–5 mm long, sessile or nearly so, apiculate; n = 108.

COMMON NAME: Dwarf Scouring Rush.

HABITAT: Moist, shaded woodlands.

RANGE: Greenland to Alaska, south to Washington, Illinois, and New York; Europe; Asia.

ILLINOIS DISTRIBUTION: Very rare; known only from Lake and McHenry Counties, and not from the latter county for nearly 100 years.

This is the most easily recognized and least variable species of *Equisetum* in Illinois, due to its small stature, curly stems, and limited number of sheath-teeth.

2. Equisetum variegatum Schleich. in Usteri, in Neue Ann. Bot. 21:120–35. 1797. *Fig. 71.*

Stems evergreen, erect, firm, unbranched above the base, slender, strongly grooved, to 25 cm tall, with 4–10 angled ridges, with each ridge bearing tubercles in two rows, with a central cavity less than one-half the diameter of the stem; sheaths green below, dark above, with 5–10 teeth, the tip of each tooth usually deciduous and with a central groove throughout; cones 5–10 mm long, apiculate; n = 108.

COMMON NAME: Variegated Scouring Rush; Variegated Horsetail.

HABITAT: Lake shores and stream banks; margins of swamps.

RANGE: Greenland to Alaska, south to California, Illinois, and Pennsylvania; Europe; Asia.

ILLINOIS DISTRIBUTION: Rare; known only from four isolated counties in the northern one-fourth of the state.

This species is distinguished from *E. × trachyodon* by its strongly grooved, 4–10 angled stem and by the deciduous teeth of its sheaths.

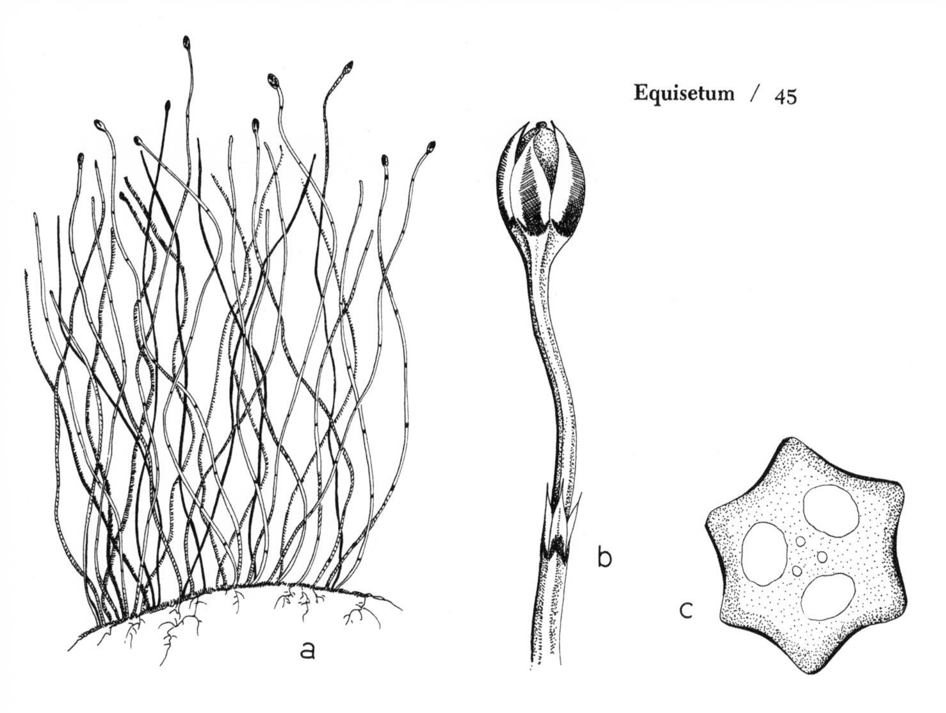

70. *Equisetum scirpoides* (Dwarf Scouring Rush). *a.* Habit, X½; *b.* Cone, stem, and node, X6; *c.* Cross-section of stem, X25.

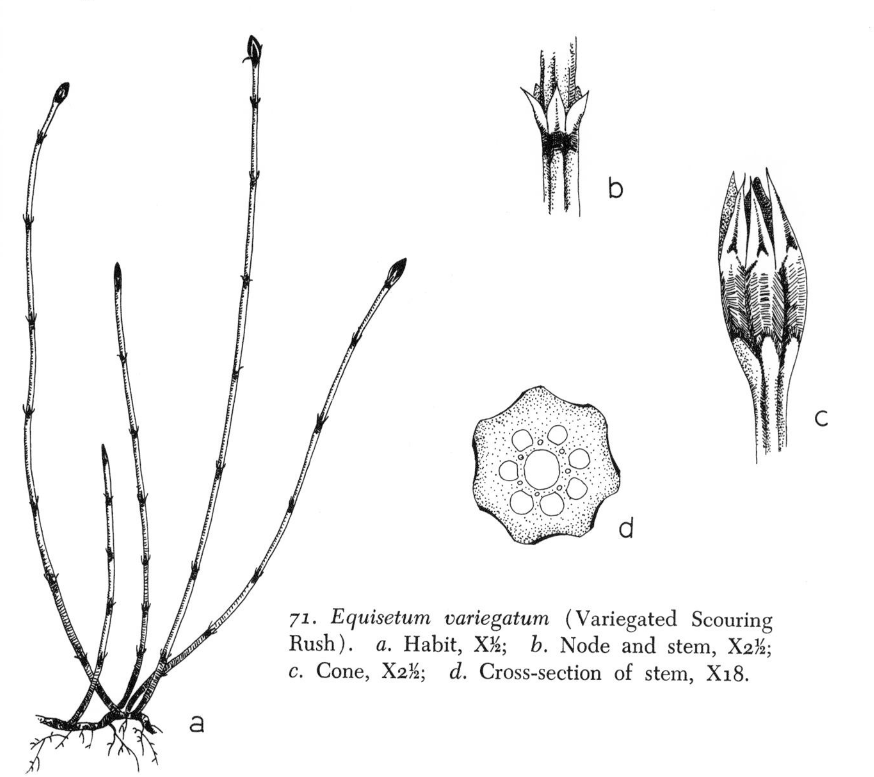

71. *Equisetum variegatum* (Variegated Scouring Rush). *a.* Habit, X½; *b.* Node and stem, X2½; *c.* Cone, X2½; *d.* Cross-section of stem, X18.

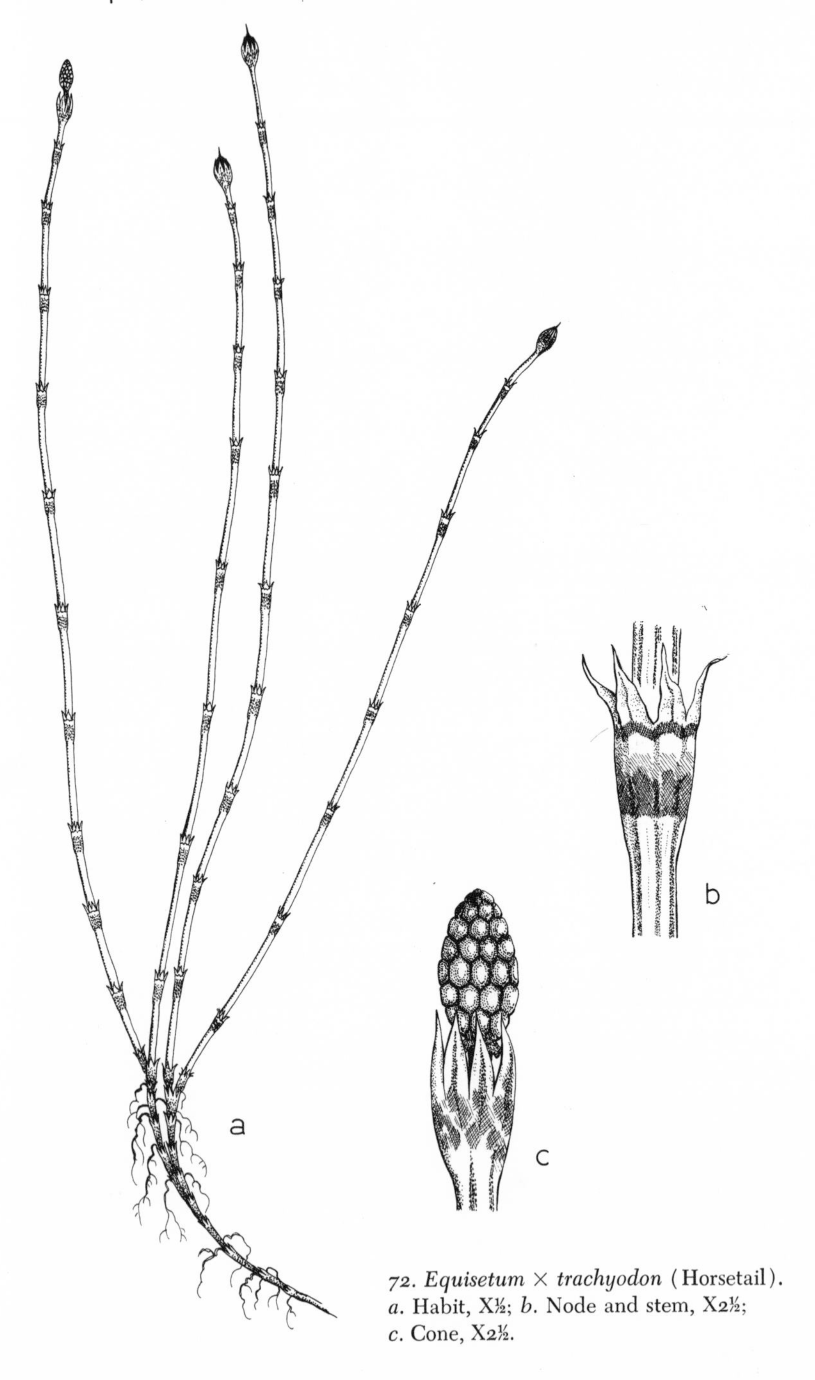

72. *Equisetum* × *trachyodon* (Horsetail). *a*. Habit, X½; *b*. Node and stem, X2½; *c*. Cone, X2½.

3. **Equisetum × trachyodon** A. Br. in Flora 20:308. 1839. *Fig. 72.*

Equisetum variegatum var. *jesupii* A. A. Eaton ex Gilb. List N. Am. Pterid. 9:27. 1901.

Stems evergreen, erect, firm, unbranched above the base, slender, shallow grooved, to 40 cm tall, with 10–17 angled ridges, with each ridge bearing tubercles in two rows, with a central cavity usually one-half to nearly two-thirds the diameter of the stem; sheaths green below, dark above, with 10–17 teeth, the tip of each tooth persistent and with a central groove throughout; cones 6–10 mm long, apiculate.

COMMON NAME: Horsetail.

HABITAT: Lake shores.

RANGE: Quebec to Michigan, south to Illinois, Indiana, and New England; Europe.

ILLINOIS DISTRIBUTION: Rare; only one specimen seen, from Cook County.

This is the plant which is sometimes known as *E. variegatum* var. *jesupii.* Hauke considers it a hybrid between *E. variegatum* and *E. hyemale.* It certainly shows numerous characters transitional between the reputed parents.

The key characters for identification of *E.* × *trachyodon* are the shallowly grooved stems with 10–17 angled ridges and the persistent teeth of the sheaths.

4. **Equisetum × nelsonii** (A. A. Eaton) Schaffner, in Am. Fern Journ. 16:45. 1926. *Fig. 73.*

Equisetum variegatum var. *nelsoni* A. A. Eaton, in Fern Bull. 12:41. 1904.

Stems annual, flexible, unbranched above the base, to 35 cm tall, with 5–12 rounded ridges, with each ridge bearing tubercles in one row, with a central cavity less than one-half the diameter of the stem; sheaths green, with 5–12 teeth, the teeth persistent and with a central groove only at tip; cones 5–10 mm long, apiculate.

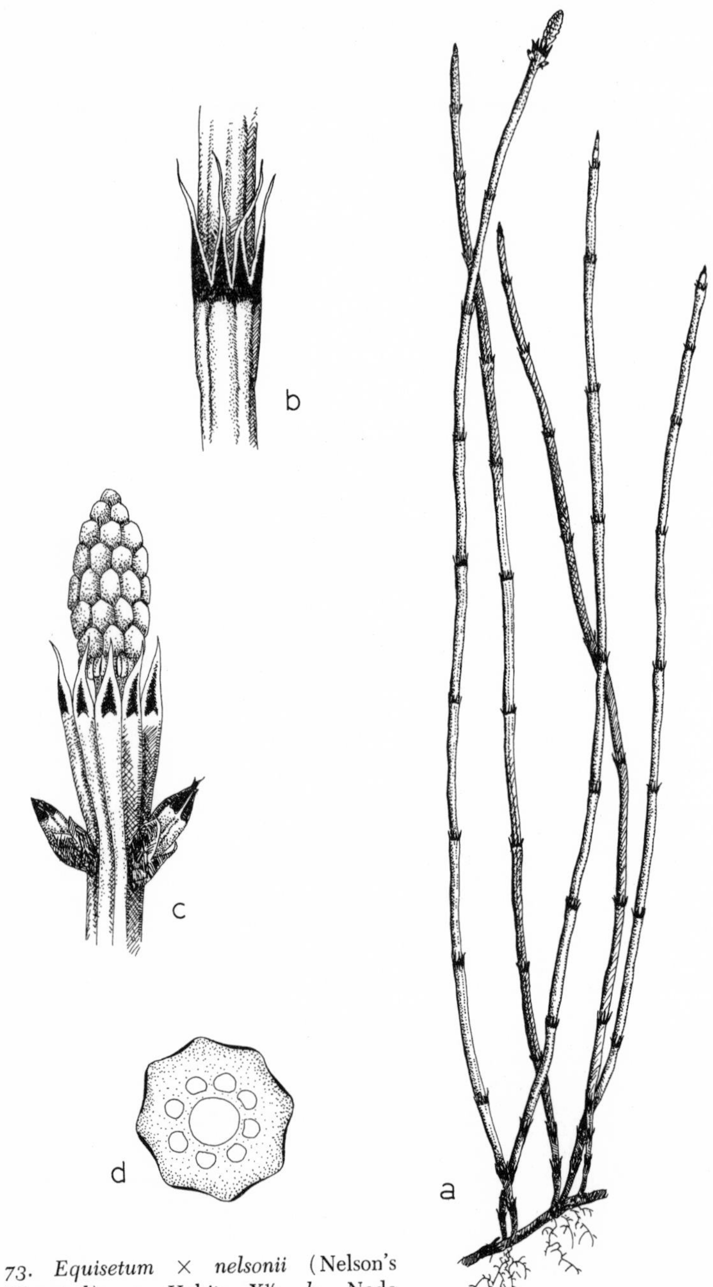

73. *Equisetum* × *nelsonii* (Nelson's Horsetail). *a.* Habit, X½; *b.* Node and stem, X7½; *c.* Cone, X2½; *d.* Cross-section of stem, X10.

COMMON NAME: Nelson's Horsetail.
HABITAT: Sandy shores.
RANGE: Labrador to Montana, south to Illinois and Connecticut.
ILLINOIS DISTRIBUTION: Rare; known only from Cook County.

This species, considered by many to be a variety of *E. variegatum*, is maintained as a distinct species here, following Schaffner's reasoning (1926). It differs from *E. variegatum* in its flexible, annual stems, the rounded ridges of the internodes with tubercles in single rows, and the teeth of the sheath with the central groove only near the tip.

Hauke believes *E.* × *nelsonii* to represent the hybrid between *E. variegatum* and *E. laevigatum*, a view followed in this work.

5. Equisetum palustre L. Sp. Pl. 1061. 1753. *Fig. 74.*

Equisetum palustre var. *americanum* Vict. Equis. du Quebec 51. 1927.

Stems annual, regularly branched above the base, to 35 cm tall, with 10–18 more or less rounded, roughened ridges, with a central cavity less than one-half the diameter of the stem; sheaths green, with 5–10 persistent, white-margined teeth with a dark central stripe; cone 10–20 mm long; n = 108.

COMMON NAME: Marsh Horsetail.
HABITAT: Banks of rivers and streams.
RANGE: Newfoundland to Alaska, south to California, Illinois, and Pennsylvania; Europe; Asia.
ILLINOIS DISTRIBUTION: Very rare; Tazewell County (not since 1891) and Kankakee County (sandy soil near Bonfield, May 23, 1963, *D. Windler 323*).

This species is recognized by its small central cavity, its 5–10 persistent teeth, and the branched stems.

6. Equisetum arvense L. Sp. Pl. 1061. 1753. *Fig. 75.*

Stems annual, of two kinds; fertile stems brown, appearing in early spring, to 25 cm tall, the sheath with basally connate teeth to 10 mm long; sterile stems regularly branched, green, appearing in late spring, to 60 cm tall, with 4–14 ridges, with the central cavity less than one-half the diameter of the stem; teeth of

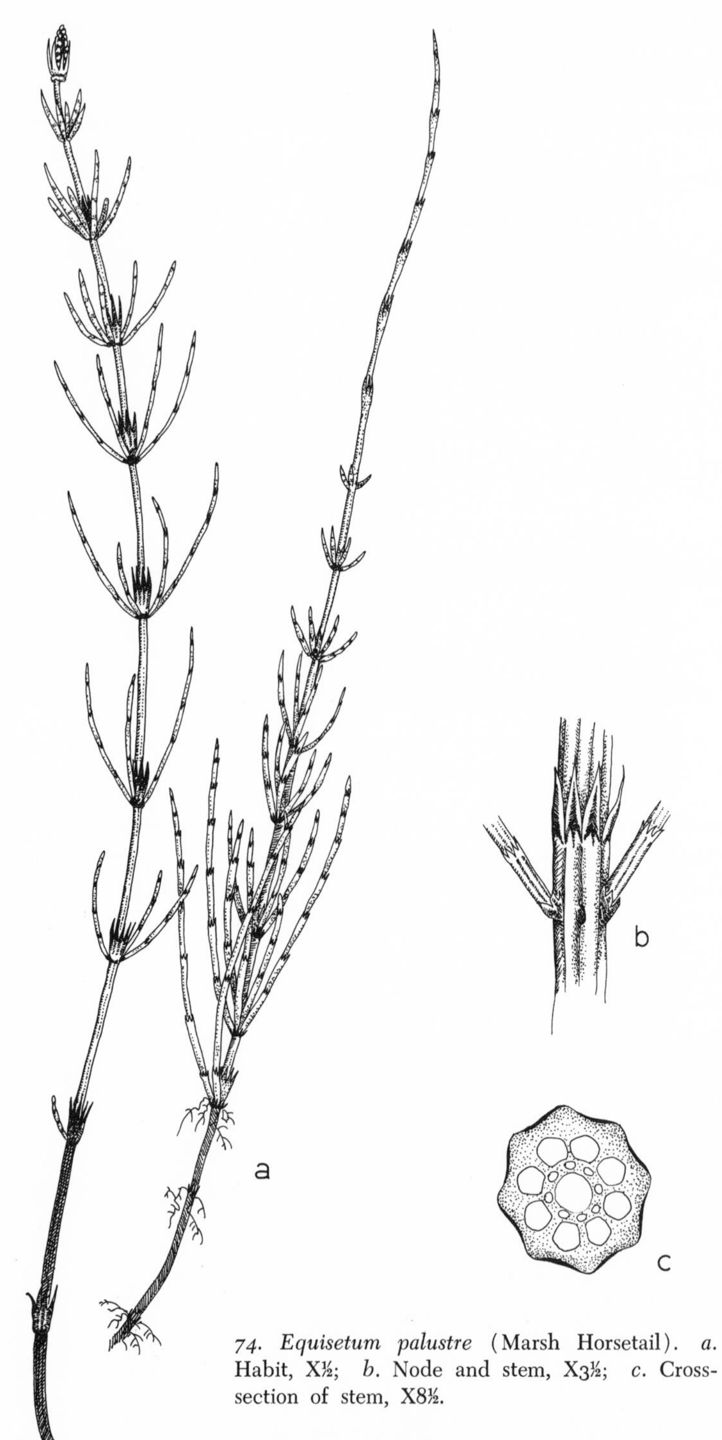

74. *Equisetum palustre* (Marsh Horsetail). *a.* Habit, X½; *b.* Node and stem, X3½; *c.* Cross-section of stem, X8½.

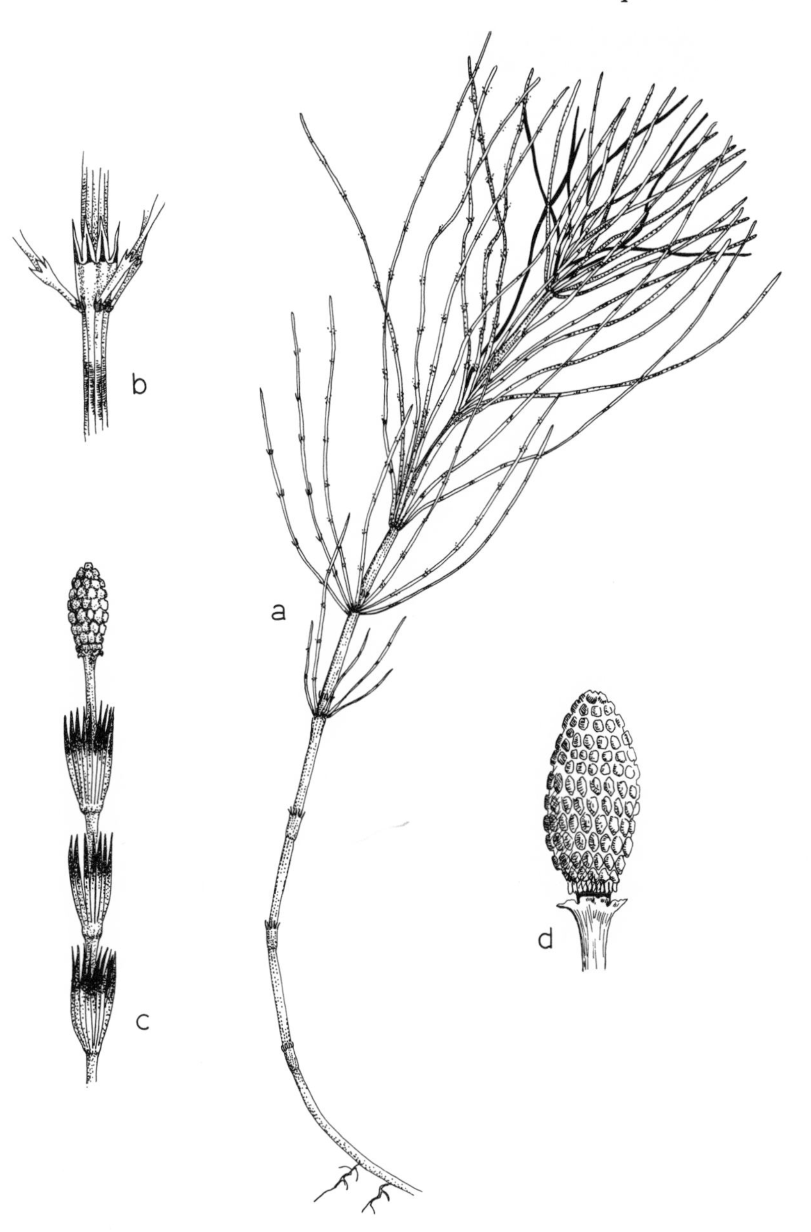

75. *Equisetum arvense* (Common Horsetail). *a.* Sterile stem, X½; *b.* Node and stem, X1½; *c.* Fertile stem, X½; *d.* Cone, X1¼.

sheaths of fertile stems 3–4, free or basally connate, to 2 mm long; cone 10–30 mm long, pedunculate, not apiculate; n = 108.

COMMON NAME: Common Horsetail.
HABITAT: Railroad embankments; roadsides; fields; shores.
RANGE: Throughout North America; Europe; Asia.
ILLINOIS DISTRIBUTION: Common; in every county.
This is one of the most variable species of the genus *Equisetum* in Illinois. The branches are usually simple, but may be compound, the stems are usually erect, but may be prostrate, and the branches may be 3- or 4-angled. While most variations have been given names, it is virtually impossible to distinguish these minor and often intergrading forms in Illinois.

7. **Equisetum × litorale** Kuhl. in Beitr. Pflanz. Russ. Reichs 4:91. 1845. *Fig. 76.*

Stems annual, erect, generally branched above the base, to 0.7 m tall, smooth, with 6–18 ridges, becoming deeply grooved near summit, with a central cavity about one-half to two-thirds the diameter of the stems; sheaths brownish or greenish, with white-margined teeth united in twos or threes; cone up to 2 cm long, not apiculate; spores abortive.

COMMON NAME: Horsetail.
HABITAT: Shores and banks.
RANGE: Quebec to British Columbia, south to Washington, Illinois, Pennsylvania, and New Jersey; Europe.
ILLINOIS DISTRIBUTION: Rare; known only from one northern county.
Equisetum × litorale is a hybrid between *E. arvense* and *E. fluviatile*. The white-margined teeth of the sheaths and the smaller central cavity of the stem distinguish *E. × litorale* from *E. fluviatile,* while *E. arvense* has an even smaller central cavity of the stem.

Plants with unbranched stems, known as *f. gracile* (Milde) Vict., may eventually be found in northern Illinois.

There is considerable variation in the number of ridges on the stem.

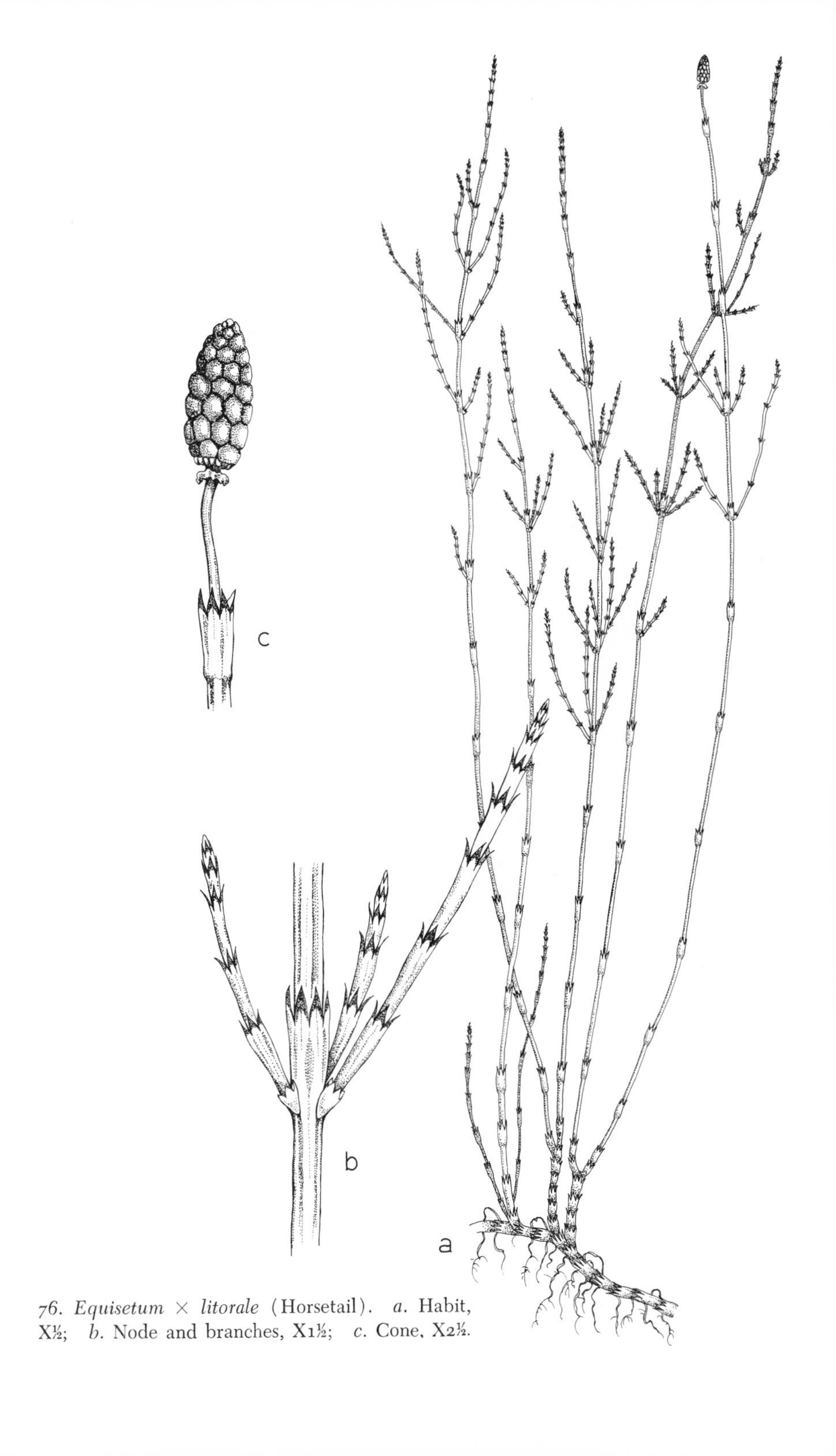

76. *Equisetum* × *litorale* (Horsetail). *a.* Habit, X½; *b.* Node and branches, X1½; *c.* Cone, X2½.

8. Equisetum hyemale L. var. **affine** (Engelm.) A. A. Eaton, in Fern Bull. 11:111. 1903. *Fig. 77.*

Equisetum praealtum Raf. Fl. Ludov. 13. 1817.

Equisetum robustum A. Br. ex Engelm. in Am. Journ. Sci. & Arts 46:88. 1844.

Equisetum robustum var. *affine* Engelm. in Am. Journ. Sci. & Arts 46:88. 1844.

Equisetum hyemale var. *robustum* (A. Br.) A. A. Eaton, in Fern Bull. 11:75. 1903.

Hippochaete prealta var. *pseudohyemalis* Farw. in Am. Fern Journ. 7:76. 1917.

Equisetum hyemale var. *pseudohyemale* (Farw.) Morton in Gleason, New Ill. Fl. No. U. S. 1:16. 1952.

Stems evergreen, erect, unbranched above the base, to sometimes over 1 m tall, more or less roughened, with up to 40 rounded ridges, with a central cavity more than two-thirds (usually three-fourths) the diameter of the stem; sheaths gray, girdled with black, with the basally connate teeth persistent or deciduous; cone variable in size, short-pedunculate, apiculate; n = 108.

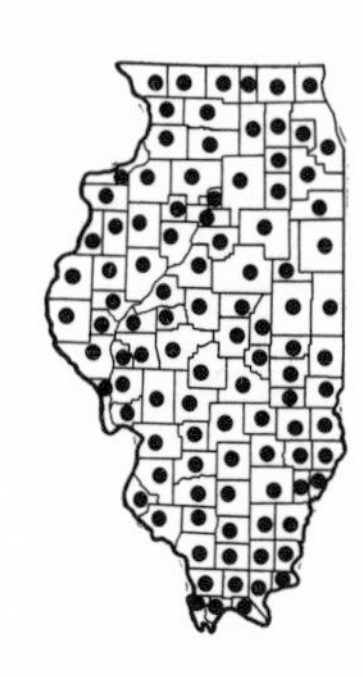

COMMON NAME: Scouring Rush.

HABITAT: Shores and banks; roadsides.

RANGE: Throughout North America; Europe; Asia.

ILLINOIS DISTRIBUTION: Common; in every county.

This *Equisetum* is the most variable one in Illinois, varying in degree of stem roughness, branching, size, and length of persistence of the teeth.

Typical var. *hyemale* is Eurasian. Our material frequently has been considered to be composed of two taxa—those (the more common ones) with more slender stems and deciduous teeth of the sheaths (var. *pseudohyemale* or var. *affine*) and those with more stout stems and rather persistent teeth of the sheaths (var. *elatum, E. praealtum,* or *E. robustum*). Since the distinguishing characters seem to be almost hopelessly overlapping, at least in Illinois material, all plants from this state are known in this treatment as *E. hyemale* var. *affine* since *affine* is the first epithet of varietal rank available.

Because this scouring rush is so abundant in Illinois, and because its morphology appears to be highly variable, a study of this variability would be desirable.

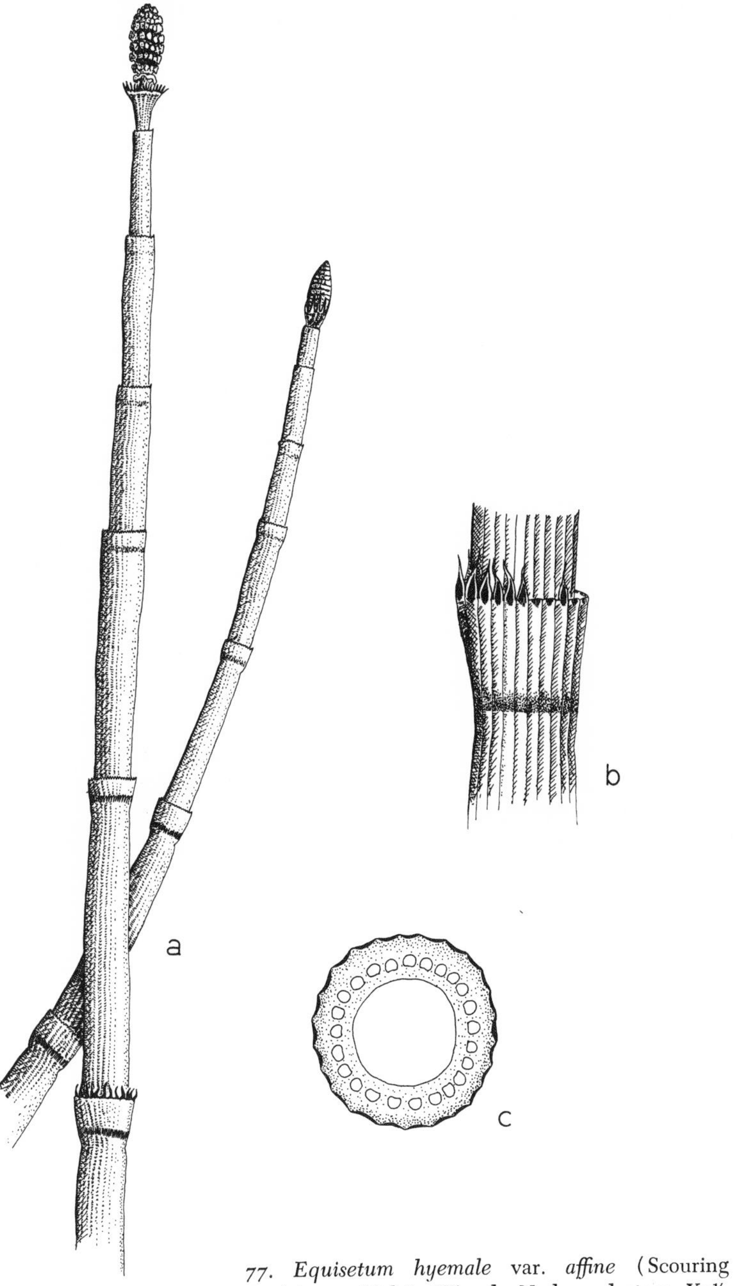

77. *Equisetum hyemale* var. *affine* (Scouring Rush). *a.* Habit, X½; *b.* Node and stem, X1¼; *c.* Cross-section of stem, X2¼.

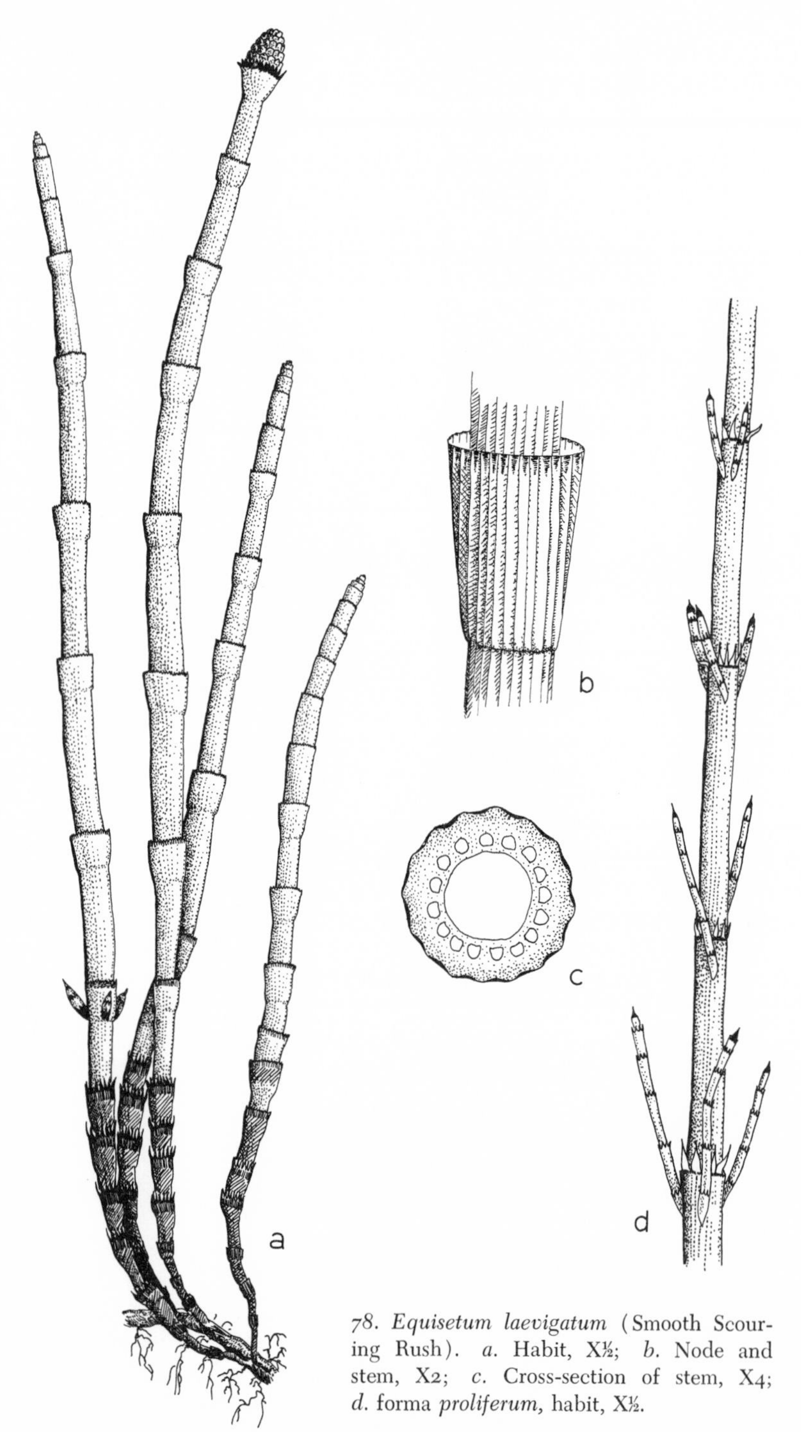

78. *Equisetum laevigatum* (Smooth Scouring Rush). *a.* Habit, X½; *b.* Node and stem, X2; *c.* Cross-section of stem, X4; *d.* forma *proliferum,* habit, X½.

9. **Equisetum laevigatum** A. Br. ex Engelm. in Am. Journ. Sci. & Arts 46:87. 1844. *Fig.* 78.

Equisetum kansanum Schaffner, in Ohio Nat. 13:21. 1912.
Equisetum laevigatum f. *proliferum* Haberer, in N. Y. State Mus. Bull. 243–44:47. 1923.

Stem annual or perennial and deciduous, erect, unbranched or rarely branched above the base, to nearly 1 m tall, more or less smooth, with up to 25 rounded ridges, with a central cavity more than one-half (usually three-fourths) the diameter of the stem; upper sheaths green, the lower green or gray, with the basally connate teeth early deciduous; cone variable in size, short-pedunculate, rounded or acute.

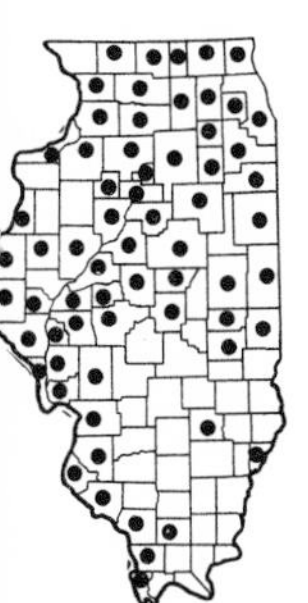

COMMON NAME: Smooth Scouring Rush.
HABITAT: Open, moist, sandy areas.
RANGE: Quebec to British Columbia, south to California, Texas, and West Virginia; Mexico.
ILLINOIS DISTRIBUTION: Common in the northern half of the state, occasional in the southern half.

This is another variable species in which there is considerable disagreement over the correct position of the variants. Several authors consider those specimens with very smooth stems and rounded cones to be a distinct species, known as *E. kansanum.* There does not seem to be complete correlation between the condition of the stem and the shape of the cone apex. Therefore, specimens which would be assigned by some workers to *E. kansanum* are considered here under *E. laevigatum.*

Forma *proliferum,* with proliferating branches from the upper nodes, has been found in several counties.

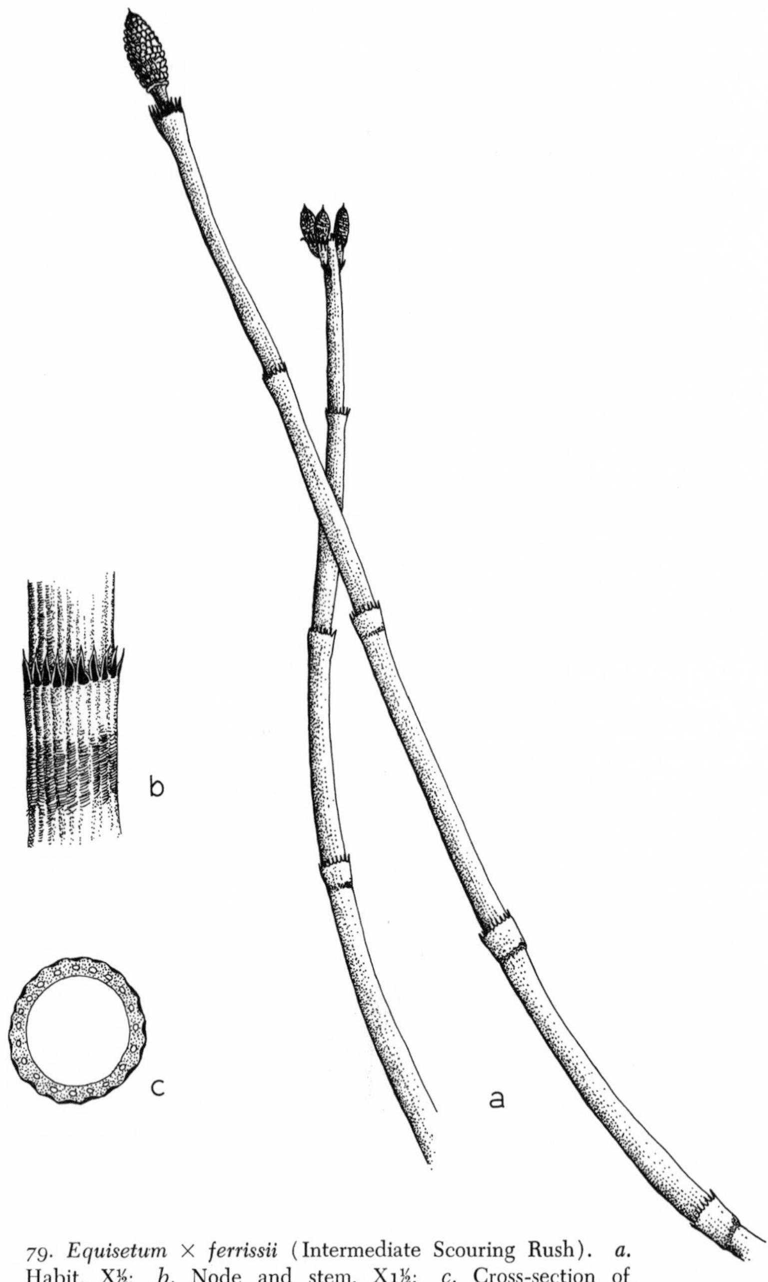

79. *Equisetum × ferrissii* (Intermediate Scouring Rush). *a.* Habit, X½; *b.* Node and stem, X1½; *c.* Cross-section of stem, X2½.

10. Equisetum × ferrissii Clute, in Fern Bull. 12:22. 1904. *Fig. 79.*

Equisetum laevigatum var. *elatum* Engelm. in Am. Journ. Arts & Sci. 46:87. 1844.

Equisetum hyemale var. *intermedium* A. A. Eaton, in Fern Bull. 10:120. 1902.

Equisetum intermedium (A. A. Eaton) Rydb. Fl. Rocky Mts. 1053. 1917.

Equisetum hyemale var. *affine* × *laevigatum* Hauke, in Am. Fern Journ. 52:126. 1962.

Stem generally evergreen, erect, unbranched above the base, to nearly 1 m tall, slightly roughened, with a central cavity more than one-half (usually four-fifths) the diameter of the stem; sheaths (at least the upper) green, with the basally connate teeth early deciduous; cone short-pedunculate, apiculate.

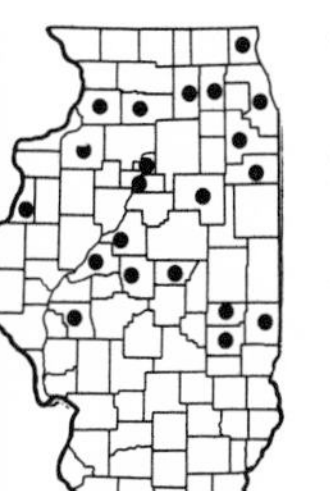

COMMON NAME: Intermediate Scouring Rush.

HABITAT: Banks and shores.

RANGE: Connecticut to British Columbia, south to California, Texas, and Virginia; Mexico.

ILLINOIS DISTRIBUTION: Scattered throughout the state, although more plentiful in the northern counties. All but one specimen examined from Illinois had previously been called *E. laevigatum* or *E. hyemale*. The roughened stems and green sheaths distinguish this hybrid.

The reasons for assuming this taxon to be a hybrid between *E. hyemale* and *E. laevigatum* are the intermediate characters present and the abortive nature of the spores. The greenish sheaths recall *E. laevigatum*, while the roughened stem is reminiscent of *E. hyemale* var. *affine*.

Hauke (1962) considers *E. laevigatum* var. *elatum* to be equivalent to *E.* × *ferrissii*.

The type locality is Will County, Illinois.

11. Equisetum fluviatile L. Sp. Pl. 1062. 1753. *Fig. 80.*

Equisetum limosum L. Sp. Pl. 1062. 1753.

Stems annual, erect, unbranched or branched above the base, to 1 m tall, smooth, with up to 25 rounded ridges, with a large central cavity about five-sixths the diameter of the stem; sheaths brownish or green, with 15–20 (–22) free, persistent teeth; cone 10–25 mm long, long-pedunculate, not apiculate.

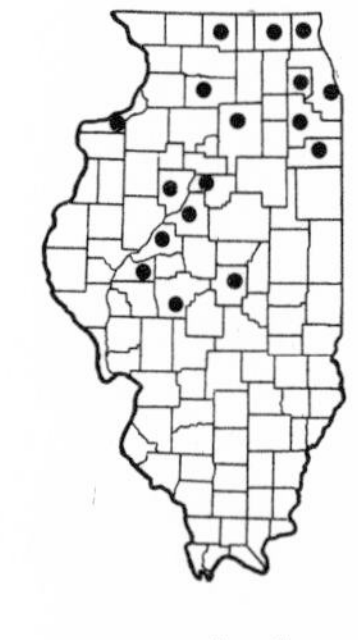

COMMON NAME: Water Horsetail.

HABITAT: In water of lakes, streams, and swamps.

RANGE: Newfoundland to Alaska, south to Oregon, Illinois, and Pennsylvania; Europe; Asia.

ILLINOIS DISTRIBUTION: Occasional in the upper half of Illinois; absent from the southern half of the state.

This is the most aquatic species of *Equisetum* in Illinois. The stem also has the largest central cavity of any species within the state. Although some variation occurs throughout the range of this species, there seems to be little variability in the Illinois material.

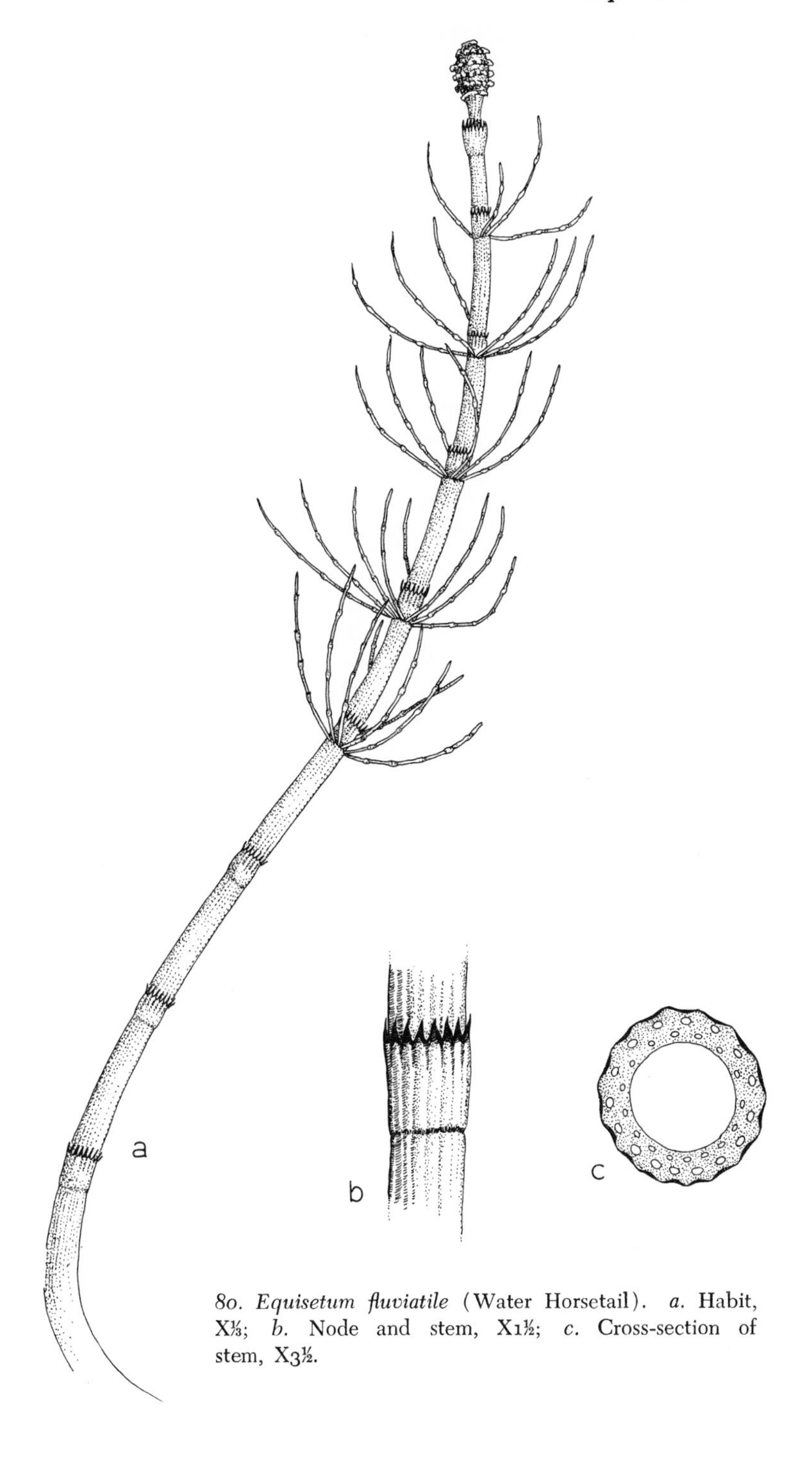

80. *Equisetum fluviatile* (Water Horsetail). *a.* Habit, X⅓; *b.* Node and stem, X1½; *c.* Cross-section of stem, X3½.

Order Ophioglossales

Only the following family comprises this order.

OPHIOGLOSSACEÆ – ADDER'S-TONGUE FAMILY

This family generally is considered to contain the most primitive ferns of temperate North America. The gametophyte generation recalls, to some extent, that of certain species of *Lycopodium* in being small, subterranean, and mycorhizal. The following description of the family is based on the sporophyte generation.

Sporophyte with small, erect, subterranean stem bearing fleshy roots; roots mycorhizal; leaf 1–2 (–3), simple to pinnately compound, reticulate-nerved, not circinate, the stipe (petiole) enclosing a basal bud; sporangia numerous, large, sessile or stipitate, 2-valved, crowded into a stipitate spike-like structure arising from the base of the blade; annulus absent.

Two genera occur in Illinois. A monograph by Clausen (1938) is the latest treatment of the family. Excellent older works by Prantl (1883; 1884) have provided the foundation for later studies.

KEY TO THE GENERA OF Ophioglossaceae IN ILLINOIS

1. Leaves pinnately compound or simple and deeply lobed, the venation free; sporangia stipitate _____________________ 1. *Botrychium*
1. Leaves simple, entire, the venation net-like; sporangia sessile ____ _______________________________________ 2. *Ophioglossum*

1. Botrychium **SW. – Grape Fern**

Perennials with fleshy roots; leaves 1–2 (–3), pinnately compound or simple and deeply lobed, frequently ternate, the venation free; sporangium-bearing spike branched, the sporangia crowded, stipitate, lateral; spores yellowish, thick-walled, not smooth.

Considerable variation exists in leaf-lobing within the genus, and taxonomists are not in agreement in their treatments of the taxa.

Of the four species known from Illinois, all but one has leaves which persist during the winter, either in a bronze or green condition.

KEY TO THE SPECIES OF Botrychium IN ILLINOIS

1. Blade of leaf long-petiolate, evergreen (often turning bronze in winter), coriaceous, subcoriaceous, or membranous.
   2. Ultimate segments of the leaf obtuse or rounded; blade fleshy or coriaceous, not bronze in winter____________1. *B. multifidum*
   2. Ultimate segments of the leaf acute or subacute; blade subcoriaceous or membranous, turning bronze in winter, or remaining green in *B. biternatum.*
      3. Pinnules entire, crenulate, or lobed; blade subcoriaceous, turning bronze in winter________________2. *B. dissectum*
      3. Pinnules sharply serrate; blade membranous, remaining green in winter_____________________________3. *B. biternatum*
1. Blade of leaf sessile, deciduous, membranous____4. *B. virginianum*

1. Botrychium multifidum (Gmel.) Rupr. ssp. **silaifolium** (Presl) Clausen, in Bull. Torrey Club 64:271. 1937. *Fig. 81.*

Botrychium silaifolium Presl, Rel. Haenk. 1:76. 1825.
Botrychium ternatum subvar. *intermedium* D. C. Eaton, Ferns N. Am. 1:149. 1878.
Botrychium ternatum var. *intermedium* D. C. Eaton in Gray, Man. 694. 1890.
Botrychium multifidum var. *intermedium* (D. C. Eaton) Farw. in Ann. Rep. Mich. Acad. Sci. 18:87. 1916.
Botrychium multifidum var. *silaifolium* (Presl) Broun, Index N. Am. Ferns 41. 1938.

Fleshy perennial to 40 cm tall, with fleshy roots; common stalk 1–5 cm long; blades ternately compound, coriaceous or fleshy when fresh, 10–20 cm long, half to two-thirds as broad, not turning bronze in autumn, the pinnules oblong to ovate, crenulate, obtuse or rounded, the stalk 3–10 cm long; fertile spike bi- or tripinnate, the stalk up to 20 cm long; sporangia globoid, less than 1 mm in diameter; spores yellowish.

COMMON NAME: Northern Grape Fern.
HABITAT: Rich woodlands.
RANGE: New Brunswick to British Columbia, south to California, northern Illinois, and New Jersey.
ILLINOIS DISTRIBUTION: Rare; in the extreme northern counties.
The typical subspecies of *B. multifidum,* which occurs north of Illinois, is a smaller plant with more crowded leaf divisions.

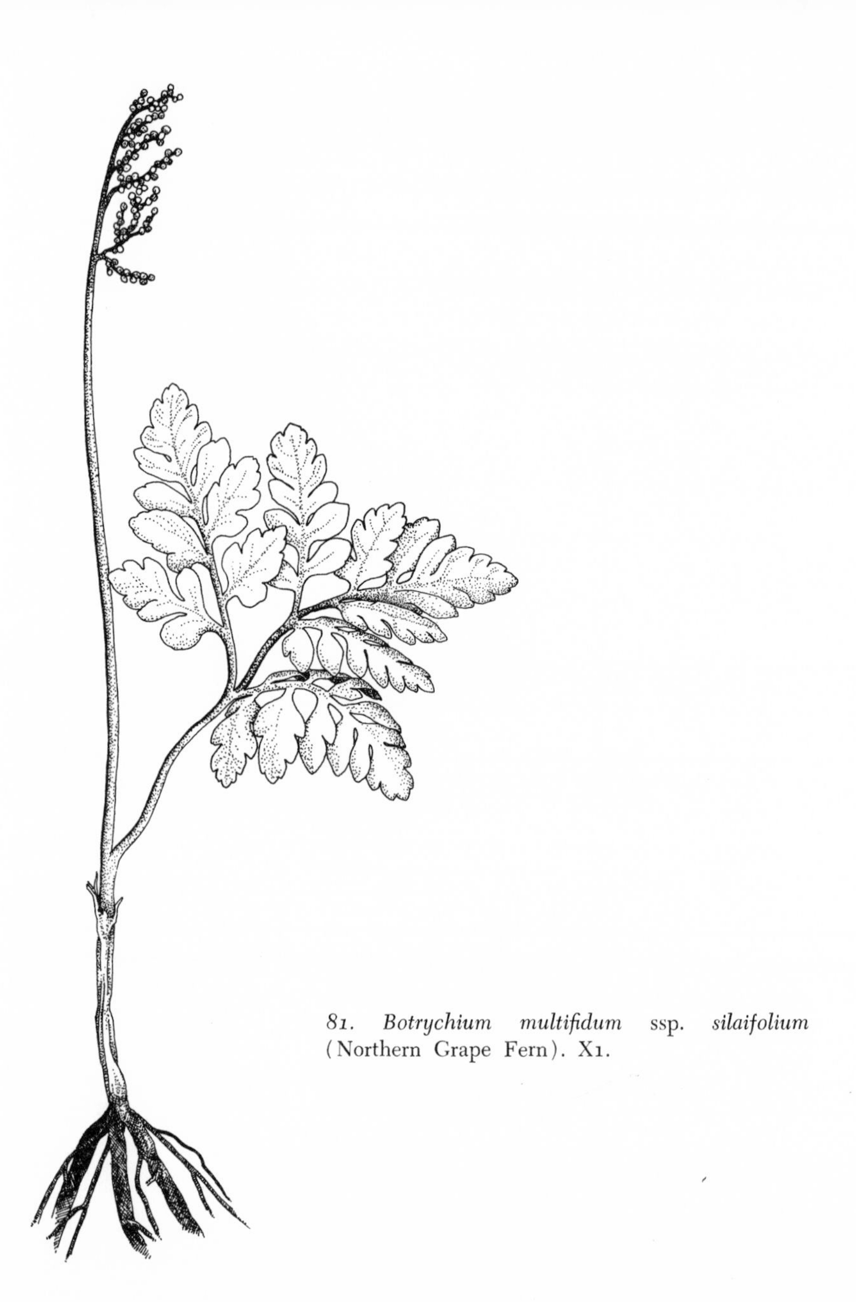

81. *Botrychium multifidum* ssp. *silaifolium* (Northern Grape Fern). X1.

Botrychium multifidum ssp. *silaifolium* may be confused with *B. dissectum* var. *obliquum*. Subspecies *silaifolium* is fleshier and bears sporangia one or two months earlier (August and September).

2. Botrychium dissectum Spreng. in Anleit. Kennt. Gewächse 3:172. 1804.

Evergreen perennial to about 30 cm tall, with fleshy roots; common stalk 4–5 cm long; blades ternately compound, subcoriaceous, 8–14 cm long, 8–15 cm broad, turning bronze in the autumn, the pinnules elongated, pinnately lobed, acute or subacute, entire or crenulate, the stalk 5–9 cm long; fertile spike bi- or tripinnate, the stalk up to 20 cm long; sporangia globoid, 0.5–1.0 mm in diameter; spores yellowish.

Two varieties, based entirely on the leaf blades, may be found in Illinois. The following key separates these taxa:

a. Blade of leaf very finely divided____2a. *B. dissectum* var. *dissectum*
a. Blade of leaf shallowly divided____2b. *B. dissectum* var. *obliquum*

2a. Botrychium dissectum Spreng. var. **dissectum** *Fig. 82.*

Botrychium lunarioides var. *dissectum* Gray, Man. 645. 1848.
Botrychium ternatum var. *dissectum* D. C. Eaton, Ferns N. Am. 1:150. 1878.
Botrychium obliquum var. *dissectum* (Spreng.) Prantl, in Jahrb. Bot. Gart. Berlin 3:342. 1884.
Botrychium obliquum f. *dissectum* (Spreng.) Clute, in Fern Bull. 10:87. 1902.
Botrychium ternatum var. *obliquum* f. *dissectum* (Spreng.) Clute, in Fern Bull. 11:116. 1903.
Botrychium dissectum var. *typicum* Clausen, in Mem. Torrey Club 19:56. 1938.

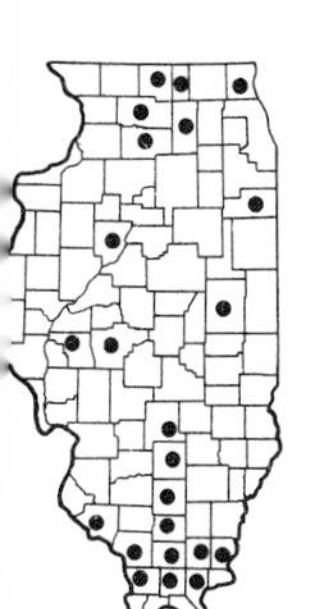

COMMON NAME: Cut-leaved Grape Fern; Bronze Fern.
HABITAT: Open oak-hickory woodlands.
RANGE: Nova Scotia to Ontario and Minnesota, south to Missouri and Florida.
ILLINOIS DISTRIBUTION: Scattered throughout the state.

Throughout most of Illinois, this variety is very distinct from var. *obliquum*. In a few counties in northeastern Illinois, specimens are of an intermediate nature and difficult to place properly.

Despite the rather large number of counties known for this variety, this plant is not frequently encountered in Illinois. It usually grows with var. *obliquum*.

82. *Botrychium dissectum* var. *dissectum* (Cut-leaved Grape Fern). X½.

83. *Botrychium dissectum* var. *obliquum* (Grape Fern). X½.

2b. Botrychium dissectum Spreng. var. **obliquum** (Muhl.) Clute, in Fern Bull. 10:76. 1902. *Fig. 83.*

Botrychium obliquum Muhl. in Willd. Sp. Pl. 5:63. 1810.
Botrychium lunarioides var. *obliquum* (Muhl.) Gray, Man. 635. 1848.
Botrychium ternatum var. *obliquum* (Muhl.) D. C. Eaton, Ferns N. Am. 1:149. 1878.
Botrychium dissectum f. *obliquum* (Muhl.) Fern. in Rhodora 23:151. 1921.

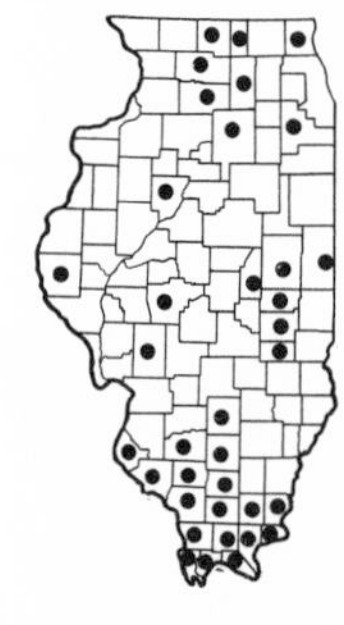

COMMON NAME: Grape Fern; Bronze Fern.
HABITAT: Open oak-hickory woodlands and pastures.
RANGE: Nova Scotia to Ontario and Michigan, south to Oklahoma and Georgia; Jamaica.
ILLINOIS DISTRIBUTION: Throughout the state; somewhat more common than var. *dissectum.*

3. Botrychium biternatum (Sav.) Underw. in Bot. Gaz. 22: 407. 1896. *Fig. 84.*

Osmunda biternata Sav. in Lam. Enc. Bot. 4:650. 1797.
Evergreen to about 35 cm tall, with fleshy or nearly fibrous roots; common stalk 5–7 cm long; blades ternately compound, membranous, to 13 cm long, to 17 cm broad, usually somewhat smaller, remaining green in winter, the pinnules elongated, generally unlobed, acute or subacute, sharply serrate; stipe to 7.5 cm long; fertile spike bi- or tripinnate, the stipe to about 15 (–25) cm long; sporangia globoid, less than 1 mm in diameter.

COMMON NAME: Grape Fern.
HABITAT: Open oak-hickory woodlands.
RANGE: Maryland to Missouri, south to Texas and Florida.
ILLINOIS DISTRIBUTION: Known only from Jackson County (near base of open hillside, Little Grand Canyon, nine miles southwest of Murphysboro, August 10, 1963, *R. H. Mohlenbrock 15150*).
This species strongly resembles *B. dissectum* var. *obliquum,* but remains green rather than bronze during

84. *Botrychium biternatum* (Grape Fern). X¾.

the winter, is smaller in all respects, and has more sharply serrate pinnules.

Most authors have considered this taxon to be a variety of *B. dissectum,* but Wherry (1961) now believes it should receive specific rank. Wagner (1961) has shown that *B. biternatum* is the correct name for this, as a species.

4. **Botrychium virginianum** (L.) Sw. in Journ. Bot. Schrad. 1800 (2):111. 1801. *Fig. 85.*

Osmunda virginiana L. Sp. Pl. 2:1064. 1753.

Botrychium virginianum var. *intermedium* Butters, in Rhodora 19:210. 1917.

Perennial to 50 (–75) cm tall, with fleshy roots; blades ternately compound, membranous, the pinnules dissected or toothed, acute or subacute, compound, membranous, to 18 (–21) cm long, to 30 (–35) cm broad, sessile, the pinnules dissected or toothed, acute or subacute; fertile spike pinnately compound, the stalk to 20 cm long; sporangia globoid, 0.5–1.0 mm in diameter; spores yellowish, shedding in June and July.

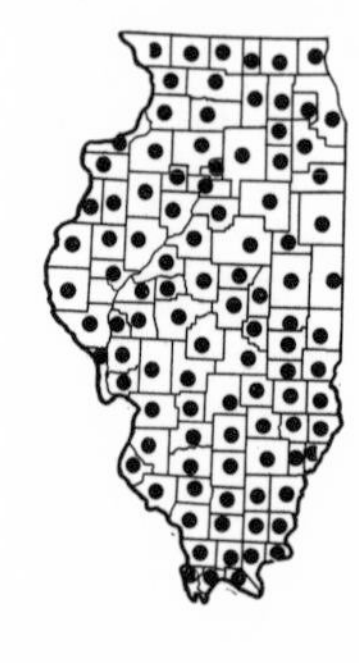

COMMON NAME: Rattlesnake Fern.
HABITAT: Dry or moist woodlands.
RANGE: Nova Scotia to British Columbia, south to California and Florida; Asia.
ILLINOIS DISTRIBUTION: Common throughout the state; known from every county.
This species is recognized readily by its sessile, deciduous leaves.

2. *Ophioglossum* [Tourn.] L. – Adder's-tongue

Perennials with fleshy roots; leaves 1–2 (–3), simple, entire, elliptic, the venation net-like; sporangium-bearing spike simple, the sporangia crowded, sessile, borne in two rows; spores yellowish, thick-walled.

This genus, because of the closed venation and simple leaf, generally is considered more advanced than *Botrychium.*

Two rather uncommon (or infrequently collected) species exist in Illinois. Both usually die back late in the summer so that they cannot be found by September.

85. *Botrychium virginianum* (Rattlesnake Fern).
a. Habit, X½; *b*, Sporangia, X5.

KEY TO THE SPECIES OF *Ophioglossum* IN ILLINOIS

1. Leaf rounded or subacute at apex, the veins forming principal areoles without secondary areoles (*Fig. 86*); plants of meadows, moist woods, or sandstone ledges ________________ 1. *O. vulgatum*
1. Leaf acute, apiculate, the veins forming principal areoles surrounding secondary areoles (*Fig. 87*); plants of limestone ledges ______ ______________________________________ 2. *O. engelmannii*

86. *Ophioglossum vulgatum* (Adder's-tongue Fern). *a.* Habit, X½; *b.* Blade, X½.

1. **Ophioglossum vulgatum** L. Sp. Pl. 1062. 1753. *Fig. 86.*

Ophioglossum arenarium E. G. Britt. in Bull. Torrey Club 24:555.1897.

Ophioglossum vulgatum f. *arenarium* (E. G. Britt.) Clute, Our Ferns in Their Haunts 316. 1901.

Ophioglossum vulgatum f. *pseudopodum* Blake, in Rhodora 15:87. 1913.

Ophioglossum vulgatum var. *pseudopodum* (Blake) Farw. in Ann. Rep. Mich. Acad. Sci. 18:84. 1916.

Ophioglossum vulgatum var. *pycnostichum* Fern. in Rhodora 41:494. 1939.

Small, rather fleshy perennial to 15 cm tall (in Illinois); blade simple, entire, elliptic to elliptic-ovate, rounded or subacute at apex, tapering to the nearly sessile or short-stipitate base, 3–6 (–8) cm long, 1–2 (–3) cm broad, glabrous, the venation composed only of primary areoles; fertile spike simple, 1–2 (–3) cm long, the sterile portion up to 10 cm long; sporangia orbicular, 0.6–1.1 mm in diameter; spores yellowish, variable; n = 250–260 (Manton, 1950).

COMMON NAME: Adder's-tongue Fern.

HABITAT: Moist woods and shaded sandstone ledges.

RANGE: Prince Edward Island to Alaska, south to Mexico; Iceland; Europe; Asia; Sao Tomé (Africa).

ILLINOIS DISTRIBUTION: Scattered in all parts of the state, except the northwestern counties. The difficulty in seeing this small fern probably accounts for its rather limited distribution.

Some botanists would restrict var. *vulgatum* to the Old World, and divide the United States specimens either into var. *pycnostichum* or var. *pseudopodum.* The differences, however, between the New and Old World forms are usually so slight and often overlapping that these divisions are unsatisfactory. Similarly, the difficulty in distinguishing var. *pycnostichum,* with a basal, membranous sheath surrounding the bud, from var. *pseudopodum,* which lacks this sheath, has influenced this author in not maintaining these varieties.

2. Ophioglossum engelmannii Prantl, in Ber. Deutsch Bot. Ges. 1:351. 1883. *Fig. 87.*

Ophioglossum vulgatum f. *engelmannii* (Prantl) Clute, Our Ferns in Their Haunts 316. 1901.

Ophioglossum vulgatum var. *engelmannii* (Prantl) Clute, Our Ferns 68. 1938.

Small, rather fleshy perennial to 15 cm tall; blades 1–3, frequently long-persisting, simple, entire, elliptic, acute and apiculate at the apex, cuneate at the base, 2.5–7.5 cm long, 1–3 cm broad, glabrous, the venation composed of primary areoles surrounding secondary areoles; fertile spike simple, 1.0–2.5 cm long, the sterile portion to 9 cm long; sporangia orbicular, 0.5–1.0 mm in diameter; spores yellowish.

COMMON NAME: Adder's-tongue Fern.

HABITAT: Limestone ledges.

RANGE: Virginia to Missouri, south to Arizona and Florida; Mexico.

ILLINOIS DISTRIBUTION: Scattered and rare in southern one-third of the state.

Ophioglossum engelmannii is distinguished from *O. vulgatum*, in fact from any other *Ophioglossum* species in the world, by the venation with primary and secondary areoles, and by the apiculate leaves. Thus far, the two Illinois species of *Ophioglossum* may be distinguished by their habitat, with *O. engelmannii* restricted to calcareous soils.

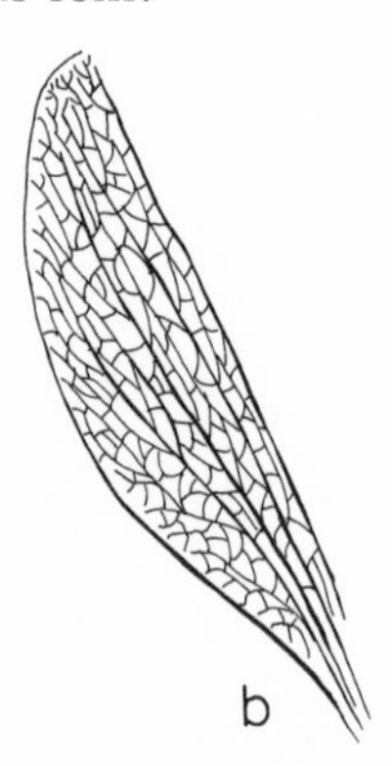

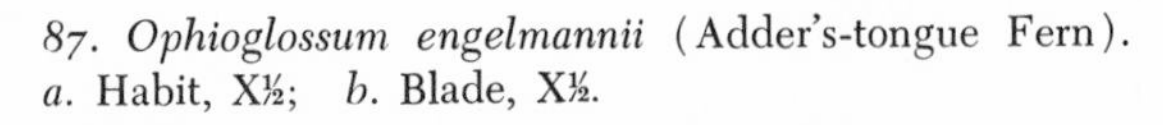
87. *Ophioglossum engelmannii* (Adder's-tongue Fern). *a.* Habit, X½; *b.* Blade, X½.

Order Filicales

This is the largest order of ferns or fern-allies in Illinois. Our representatives are of three families—the Hymenophyllaceae, or filmy ferns, the Osmundaceae, or royal ferns, and the Polypodiaceae, or true ferns.

The Hymenophyllaceae is represented in Illinois only by a single, rare species of the genus *Trichomanes.* The Osmundaceae contains a single genus with three species in Illinois. The Polypodiaceae, which is the largest family of ferns or fern-allies in the state, is composed of forty-six species distributed among eighteen genera in Illinois.

KEY TO THE FAMILIES OF Filicales IN ILLINOIS

1. Blades delicate, translucent, only a single layer of cells thick; sporangia borne at the base of a bristle-like projection from the margin of the blade (*Fig. 88b*)____________Hymenophyllaceae, page 75
1. Blades more firm, opaque, several layers of cells thick; sporangia not associated with a marginal bristle.
   2. Fertile and sterile portions of leaves essentially alike in form, except for a possible size differentiation__Polypodiaceae, page 84
   2. Fertile and sterile portions of leaves greatly dissimilar in form.
      3. Sporangia naked and solitary (although borne in loose panicles or clusters) on modified pinnae, these borne at the apex of the leaf (*Fig. 89a*), or centrally (*Fig. 91a*), or the whole leaf fertile______________________Osmundaceae, page 76
      3. Sporangia covered by an indusium or, if not, then borne in well defined sori________________Polypodiaceae, page 84

HYMENOPHYLLACEÆ—FILMY FERN FAMILY

Only the following genus occurs in Illinois.

1. Trichomanes L.—Filmy Fern

Rhizomes slender, scaleless; leaves delicate, very thin, the blade only one cell layer thick, without stomates; indusia marginal, tubular, with a bristle-like receptacle, the sporangia borne near the base of the receptacle.

Only the following species occurs in Illinois, although *T. petersii* has been erroneously attributed to Illinois by several authors.

1. Trichomanes boschianum Sturm ex Bosch, in Nederl. Kruidk. Arch. 5 (2):160. 1861. *Fig. 88.*

Rhizomes slender, wiry, black, densely hairy at first, glabrous at maturity; blades to 18 cm long, usually smaller, delicate, translucent, one cell layer thick, bipinnatifid, the pinnules more or less obtusely lobed, glabrous or sparsely hairy; petiole glabrous or sparsely hairy, winged nearly to base; sori about 2 mm long, tubular, borne along the margins of the lobes of the pinnules; n = 72.

COMMON NAME: Filmy Fern.

HABITAT: Beneath moist, overhanging sandstone cliffs.

RANGE: Ohio to Illinois, south to Alabama and Georgia; Arkansas.

ILLINOIS DISTRIBUTION: Rare; confined to Pope and Johnson Counties in extreme southern Illinois.

For a detailed discussion of this species, see Mohlenbrock and Voigt (1959) and Evers (1961).

Wherry (1961) considers this species to be only a smaller variety of the subtropical *T. radicans.*

OSMUNDACEÆ – ROYAL FERN FAMILY

Only the following genus occurs in Illinois.

1. Osmunda [Tourn.] L. – Royal Fern

Large, perennial ferns with fibrous roots; blades dimorphic, the long, scaleless petiole winged at the very base; fertile leaves (or parts of them) not leaf-like; sporangia large, stipitate, pear-shaped, bivalved; annulus poorly developed; spores numerous, green.

The three species of *Osmunda* in Illinois are among the most handsome ferns in the state. They all may attain a stature in excess of one meter. Sporangial color varies with degree of maturity.

There have been no recent studies on the taxonomy of the northeastern North American species.

Although the three Illinois species usually occur in low, swampy woodlands, they generally occupy sandstone ledges in the Shawnee Hills of southern Illinois.

88. *Trichomanes boschianum* (Filmy Fern). *a.* Habit, X1; *b.* Pinna with sorus, X4.

KEY TO THE SPECIES OF **Osmunda** IN ILLINOIS

1. Leaves bipinnate, the pinnules serrulate; sporangia borne on upper half of leaf------------------------------------1. *O. regalis*
1. Leaves always pinnate-pinnatifid, the pinnules entire; sporangia borne on a separate fertile leaf or centrally on the leaf.
   2. Sporangia borne on a separate fertile leaf----2. *O. cinnamomea*
   2. Sporangia produced centrally between sterile pinnae-------- ---------------------------------------3. *O. claytoniana*

1. Osmunda regalis L. var. **spectabilis** (Willd.) Gray, Man. 600. 1856. *Fig. 89.*

Osmunda spectabilis Willd. Sp. Pl. 5:98. 1810.

Erect, handsome perennial to nearly 1 m tall; sterile blades narrowly ovate, bipinnate, with 2–8 pairs of pinnae, each pinna bearing up to 21 pinnules, the pinnules alternate, oblong to lanceolate, subacute or obtuse at apex, asymmetrical at the rounded base, serrulate, glabrous, to 5 cm long, to 2.5 cm broad; petiolules about 1 mm long; fertile portions on upper part of leaf strongly contracted, with densely crowded, yellow-brown sporangia; n = 22 (Manton, 1950).

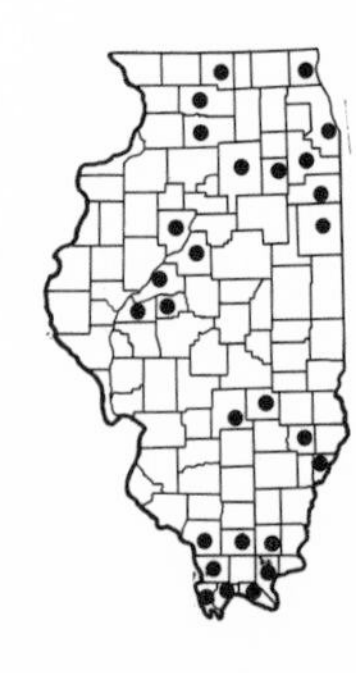

COMMON NAME: Royal Fern; Regal Fern.

HABITAT: Swamps, woods, or, in some of the southern counties, moist, sandstone ledges.

RANGE: Newfoundland to Manitoba, south to Texas and Florida; Mexico; West Indies; South America; Europe; Asia.

ILLINOIS DISTRIBUTION: Throughout the state, except the east-central area, although not particularly common.

2. Osmunda cinnamomea L. Sp. Pl. 1066. 1753. *Fig. 90.*

Erect perennial about 1 m tall from dark brown rootstocks; sterile blades lanceolate to elliptic, pinnate-pinnatifid, with up to 30 pairs of pinnae, each pinna with up to 25 pairs of lobes, without serrulations, glabrous; petiole, rachis, midvein, and leaf axils lanate, the wool reddish-brown; fertile leaves borne separately in the center of a ring of sterile leaves, much reduced, erect, the pinnules bearing many crowded brown sporangia; n = 44 (Manton, 1950).

89. *Osmunda regalis* var. *spectabilis* (Royal Fern). *a.* Habit, X¼; *b.* Pinna, X1¼; *c.* Sporangia, X1½.

90. *Osmunda cinnamomea* (Cinnamon Fern). *a.* Habit, X½; *b.* Pinna, X3; *c.* Sporangia, X1½.

COMMON NAME: Cinnamon Fern.
HABITAT: Swamps and swampy woods or rarely on moist, sandstone ledges.
RANGE: Newfoundland to Manitoba, south to New Mexico and Florida.
ILLINOIS DISTRIBUTION: In the northern one-third of the state; also Pope County (sandstone ledge, Hayes Creek Canyon, one mile west of Eddyville, *J. R. Dixon*) in the extreme southeastern part of Illinois.

Some variation occurs in the shape of the sterile blade, but this is not worthy of recognition by name. In sterile condition, the cinnamon fern is difficult to distinguish from the interrupted fern. The cinnamon fern is more densely woolly, and the pinnules are acuminate rather than obtuse.

3. **Osmunda claytoniana** L. Sp. Pl. 1066. 1753. *Fig. 91.*

Osmunda interrupta Michx. Fl. Bor. Am. 2:273. 1803.

Erect perennial to about 1 m tall; outer series of leaves sterile, broadest near the middle, with up to 26 pairs of pinnae, each pinna with up to 20 pairs of lobes, without serrulations, glabrous; inner series of leaves either all sterile, or with fertile pinnae midway between lower and upper sterile pinnae; petiole, rachis, and leaf axils usually woolly at first, becoming nearly glabrous at maturity; fertile pinnae 1–6 pairs, to 3.5 cm long, bearing numerous, globoid, sessile or subsessile, brown sporangia; n = 22 (Manton, 1950).

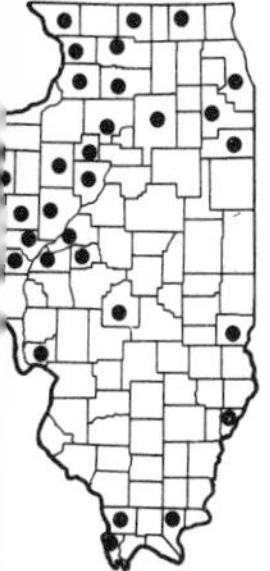

COMMON NAME: Interrupted Fern.
HABITAT: Moist, low woodlands, or rarely in moist depressions along sandstone ledges.
RANGE: Newfoundland to Manitoba, south to Nebraska, Arkansas, and Georgia.
ILLINOIS DISTRIBUTION: Occasional in the northern half of Illinois, rare in the southern half. Its thirty-two county records make it the most widely occurring *Osmunda* in Illinois.

There is little variation in the Illinois material of this species, although occasionally half-sterile, half-fertile pinnae may be encountered.

The presence of the small, grape-like sporangial clusters midway between the lower and upper sterile pinnae gives this fern its common name. Generally, there are three pairs of sterile pinnae below the fertile ones, and up to 19 pairs above the fertile ones.

In almost every Illinois specimen, sparse woolliness is present on the petioles, rachises, and in the leaf axils, although much less than in *O. cinnamomea*. It is the limited hairiness of *O. claytoniana* that best distinguishes sterile *O. claytoniana* from sterile *O. cinnamomea*.

91. *Osmunda claytoniana* (Interrupted Fern). *a.* Habit, X½; *b.* Pinna, X3; *c.* Sporangia, X2.

POLYPODIACEÆ – TRUE FERN FAMILY

Rhizomes subterranean or superficial, bearing scales and hairs; leaves simple to variously compound; fertile and sterile leaves similar or dissimilar; sporangia grouped into clusters (sori), either marginal or on the undersurface of the blade, frequently covered by an indusium.

The Polypodiaceae is the largest family in Illinois of those plants considered to be ferns or fern-allies. Eighteen genera occur in Illinois, exhibiting great diversity both in characteristics and habitats. *Phyllitis,* the hart's-tongue fern, is the only genus of Polypodiaceae in the northeastern United States not represented in Illinois. Some of the most beautiful plants in Illinois belong to this family.

Some recent workers have proposed dividing the Polypodiaceae into several families.

KEY TO THE GENERA OF Polypodiaceae IN ILLINOIS

1. Sori on the leaf margin, covered by a recurved outgrowth of the blade.
    2. Sori distinct, borne on the veins or vein tips.
        3. Leaves distinctly hairy or glandular, the petiole unbranched.
            4. Blades glandular-pubescent; rhizome long-creeping, pubescent__________________________1. *Dennstaedtia*
            4. Blades eglandular; rhizomes compact, scaly_6. *Cheilanthes*
        3. Leaves glabrous; petiole branched at the apex__2. *Adiantum*
    2. Sori continuous.
        5. Rhizome without scales; leaves ternately pinnate; coarse ferns at least 50 cm tall, mostly in woods or in fields__3. *Pteridium*
        5. Rhizome scaly; leaves regularly pinnate; ferns rarely over 35 cm tall, mostly on rocks.
            6. Stipe green, weak__________________4. *Cryptogramma*
            6. Stipe dark brown or purple-brown, wiry______5. *Pellaea*
1. Sori not marginal, or if so, not covered by leaf margin.
    7. Indusium absent.
        8. Blades once-pinnatifid; leaves less than 40 cm long, usually less than 25 cm long_____________________7. *Polypodium*
        8. Blades bipinnatifid or pinnately divided; leaves over 40 cm long.
            9. Rachis unwinged; leaves ternate_____11. *Gymnocarpium*
            9. Rachis winged (except sometimes between the lowest pinnae); leaves triangular, but not ternate____________

---------------------------------12. *Thelypteris*

7. Indusium present, although sometimes so deeply cleft as to be difficult to observe, and sometimes soon deciduous.
    10. Leaves strongly dimorphic (i.e., with the fertile quite different from the sterile).
        11. Fertile leaves bipinnate, the sterile ones pinnatifid ------------------------------------9. *Onoclea*
        11. Fertile and sterile leaves both once-pinnate---------- ---------------------------------10. *Matteuccia*
    10. Leaves all alike. (In *Polystichum,* the upper fertile pinnae are strongly modified; in one *Asplenium,* the leaves are slightly dimorphic.)
        12. Indusium attached at the center or the side of the sorus.
            13. Indusium attached at the center of the sorus.
                14. Indusium peltate; fertile pinnae apical and modified; leaf once-pinnate----8. *Polystichum*
                14. Indusium reniform; fertile pinnae identical to sterile pinnae; leaf pinnate-pinnatifid to tripinnate.
                    15. Rhizome slender, cord-like; leaves deciduous; petiole without scales, or with scales at most only 5 mm long--------- ----------------------12. *Thelypteris*
                    15. Rhizome stout; leaves more or less evergreen; petiole scaly (at least near base), the scales over 5 mm long--13. *Dryopteris*
            13. Indusium attached laterally.
                16. Sori parallel to the main vein, appearing to be in a chain----------------14. *Woodwardia*
                16. Sori parallel to the lateral veins.
                    17. Veins reaching the margin; blades deciduous------------------15. *Athyrium*
                    17. Veins not reaching the margin; blades usually evergreen--------16. *Asplenium*
        12. Indusium attached beneath the sporangia (in *Cystopteris,* the indusium may sometimes appear to be lateral).
            18. Indusium separating into shreds; petioles rather densely chaffy--------------------17. *Woodsia*
            18. Indusium hood-like, not separating into shreds; petioles sparsely chaffy, or glabrous----------- -----------------------------18. *Cystopteris*

1. *Dennstaedtia* BERNH. – Hay-scented Fern

Rhizomes hairy, scaleless; leaves bipinnate-pinnatifid, glandular-pubescent; sori marginal, distinct, covered by a recurved tooth of the leaf margin.

This genus is distinguished by its distinct, marginal, globose sori and its characteristic hay-scented odor. Only the following species occurs in Illinois. For a technical discussion of the genus, see Tryon (1960).

1. **Dennstaedtia punctilobula** (Michx.) Moore, Ind. Fil. 307. 1857. *Fig.* 92.

Nephrodium punctilobulum Michx. Fl. Bor. Am. 2:268. 1803.
Dicksonia pilosiuscula Willd. Enum. Pl. Hort. Berol. 1076. 1809.
Dicksonia punctilobula (Michx.) Gray, Man. 628. 1848.

Tall, arching, deciduous perennial to about 70 cm; rootstocks sparsely hairy, scaleless, dark brown; blades thin, deciduous, generally lanceolate, acuminate, bipinnate-pinnatifid, with up to more than 30 pairs of pinnae and up to more than 20 pairs of pinnules, the ultimate segments sessile or subsessile, obtuse, toothed, glandular-hairy beneath; petiole with sparse glandular pubescence, brownish or reddish-brown, to about 20 cm long; rachis reddish-brown below, passing through stramineous to finally green at the apex, glandular-hairy; sori marginal, terminating the veins, covered by a membranous, white indusium modified from a recurved tooth of the leaf margin; spores yellowish-brown.

COMMON NAME: Hay-scented Fern.
HABITAT: Moist, shaded sandstone ravines.
RANGE: Nova Scotia to Minnesota, south to Arkansas and Georgia.
ILLINOIS DISTRIBUTION: Rare; only in extreme southern Illinois (Pope County, along deeply shaded sandstone cliff, Lusk Creek, east of Eddyville, August 7, 1952, *W. M. Bailey and J. R. Swayne* 2759). The report from Wabash County could not be verified.

This is one of the most beautiful, as well as one of the rarest, ferns in Illinois. The gracefully delicate leaves, growing along a moist, shady, moss-covered sandstone cliff are, indeed, a magnificent masterpiece of nature.

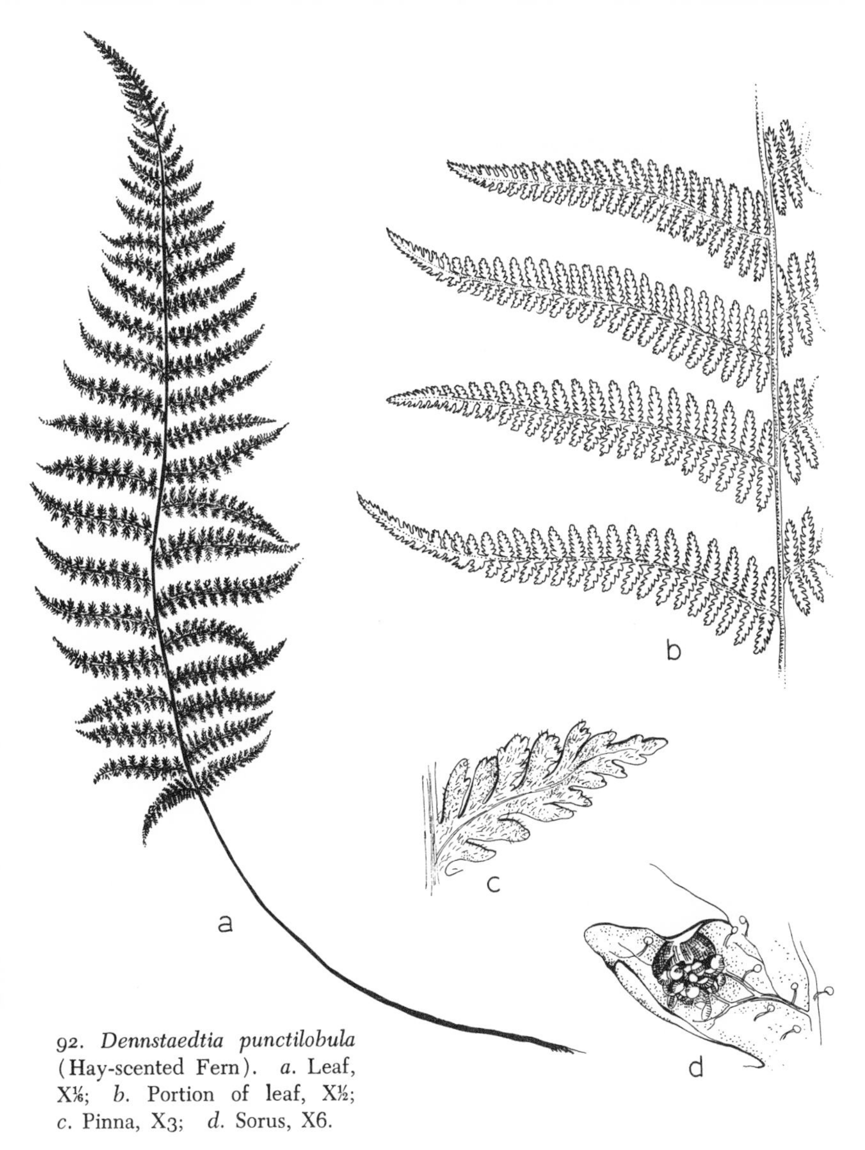

92. *Dennstaedtia punctilobula* (Hay-scented Fern). *a.* Leaf, X⅙; *b.* Portion of leaf, X½; *c.* Pinna, X3; *d.* Sorus, X6.

In sterile condition, the hay-scented fern might be confused with *Athyrium filix-femina* var. *asplenioides,* which lacks glandular hairs, with *Dryopteris intermedia* and *Woodsia obtusa,* both of which have scaly petioles, and with *Cystopteris bulbifera,* which has veins running to the sinuses, rather than to the marginal teeth. In addition, all the above species have short, thick rhizomes.

A similar species, with slender, horizontal rhizomes, is *Cystopteris fragilis* var. *protrusa,* a plant distinguished by its smaller stature and glabrous leaves.

2. *Adiantum* L. – Maidenhair Fern

Rhizomes creeping; leaves bipinnate; petiole shining, black, divided at apex; sori marginal, separate, covered by the reflexed leaf margin.

Only the following species occurs in Illinois.

1. Adiantum pedatum [Tourn.] L. Sp. Pl. 1095. 1753. *Fig. 93.* Delicate, deciduous perennial, with grayish rhizomes; petiole to 45 cm long, shiny, black or deep brown, glabrous, forked at apex into eventually several pinnate pinnae, giving the entire blade a fan-shaped appearance; pinnules up to 25 pairs, alternate, flabellate to reniform, notched along the upper margin, glabrous, petiolulate, the main vein paralleling the lower margin; lobes of upper leaf margin reflexed to cover the reniform to linear sori; spores yellowish.

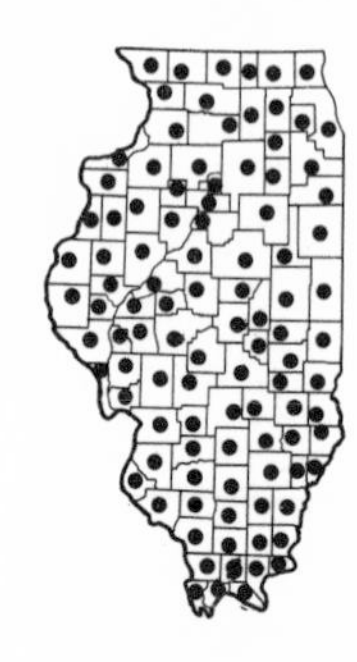

COMMON NAME: Maidenhair Fern.
HABITAT: Moist, shaded woodlands.
RANGE: Quebec to Alaska, south to California, Oklahoma, and Georgia; Asia.
ILLINOIS DISTRIBUTION: Common; in every county.

This is one of the most familiar and graceful ferns in Illinois. It occurs commonly in moist, shaded woodlands. Considerable variability exists in the leaf. Pinnule shape ranges from nearly triangular to elongate-rectangular. The terminal and lowest pinnules vary the greatest. Size of the leaf in Illinois measures from about 5 inches across (in some northern specimens) to about 9 inches across (in some southern specimens).

3. *Pteridium* GLED. – Bracken Fern

Rhizome stout, creeping, scaleless; leaves ternate, bipinnate-pinnatifid; sori marginal, continuous, covered by the reflexed leaf margin.

For a technical discussion of the genus, see Tryon (1941).

Only the following species occurs in Illinois. When collecting this species, care should be taken to gather the growing tip of the extensively creeping rhizome.

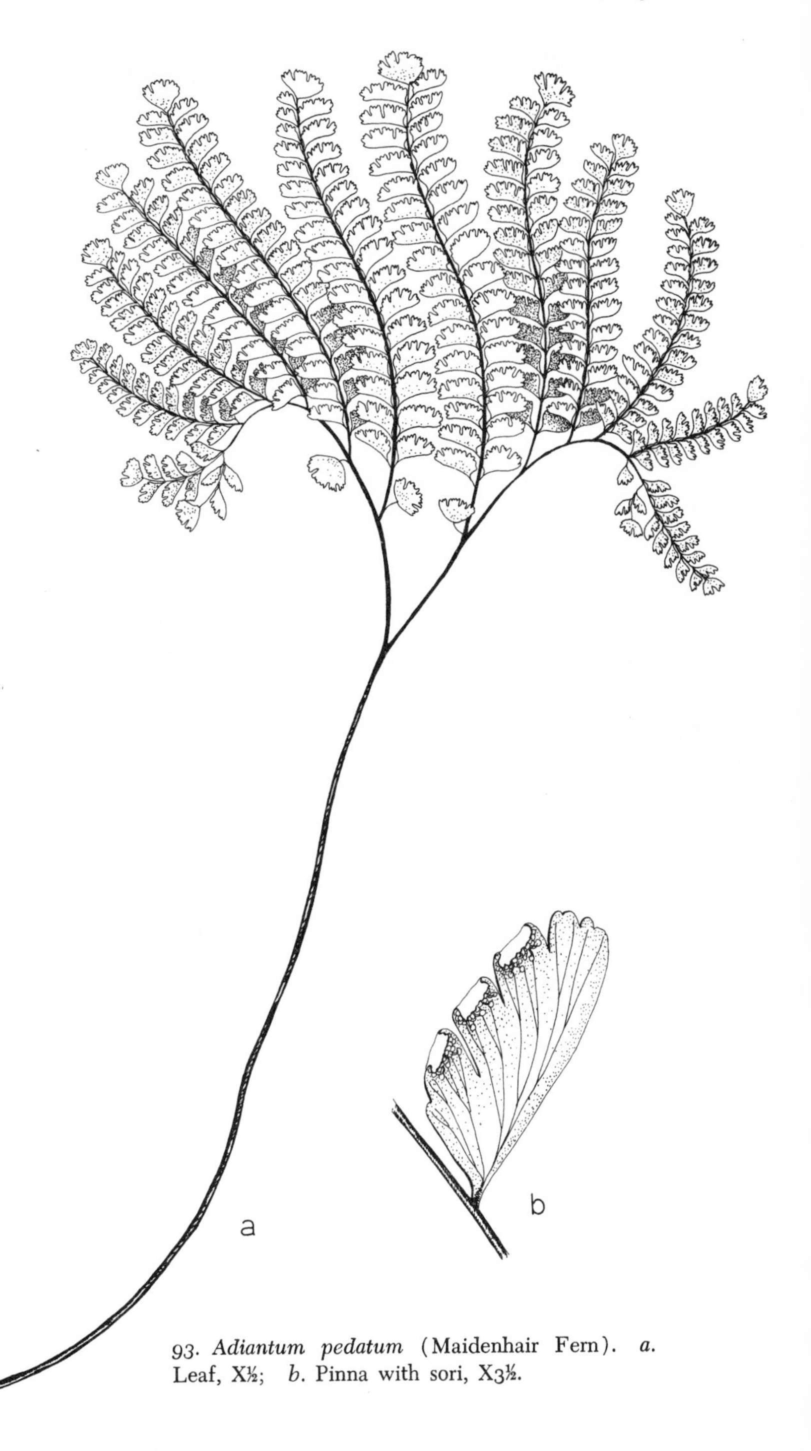

93. *Adiantum pedatum* (Maidenhair Fern). *a.* Leaf, X½; *b.* Pinna with sori, X3½.

1. **Pteridium aquilinum** (L.) Kuhn in Decken, in Reisen Ost-Afr. 3 (3):11. 1879.

Pteris aquilina L. Sp. Pl. 1075. 1753.

Asplenium aquilinum (L.) Bernh. in Journ. Bot. Schrad. 1799 (1):310. 1799.

Rhizome extensively creeping, hairy but scaleless; leaves large, sometimes to 1 m tall, the blades triangular, ternate, to about 45 cm long, bipinnate-pinnatifid, glabrous or puberulent; petioles stout, glabrous or pubescent near base; sori marginal, continuous, covered by the reflexed leaf margin.

This widely distributed fern has two rather well-marked varieties in Illinois. The typical variety, with ciliate indusia, is restricted to Europe, Asia, and Africa.

The following key separates the varieties of *P. aquilinum* in Illinois.

a. Margins of ultimate leaf segments hairy, the terminal segments 5–8 mm broad_________________1a. *P aquilinum* var. *latiusculum*

a. Margins of ultimate leaf segments glabrous or nearly so, the terminal segments to 4.5 mm broad__1b. *P. aquilinum* var. *pseudocaudatum*

1a. **Pteridium aquilinum** (L.) Kuhn var. **latiusculum** (Desv.) Underw. ex Heller, Cat. N. Am. Pl. 17. 1909. *Fig. 94a, b.*

Pteris lanuginosa Spreng. in Nova Acta 10:231. 1821.

Pteris sprengelii Steud. Nom. Bot. 2:358. 1824.

Pteris latiuscula Desv. in Mem. Soc. Linn. 6 (2):303. 1827.

Pteridium latiusculum (Desv.) Hieron. ex Fries, in Wiss. Ergebn. Schwed. Rhodesia-Kongo Exp. 1 (1):7. 1914.

Growing tip of rhizome glabrous or sparsely white-hairy; margin of ultimate leaf segments pubescent, 5–8 mm broad; n = 52 (Manton, 1950).

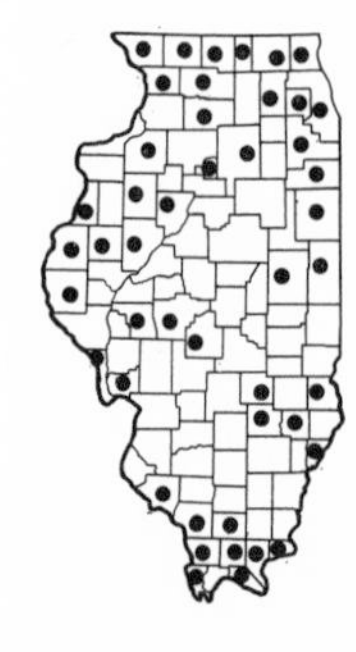

COMMON NAME: Bracken Fern.

HABITAT: Open woodlands and fields.

RANGE: Newfoundland to South Dakota, south to Colorado, Mississippi, and North Carolina; Mexico; Europe; Asia.

ILLINOIS DISTRIBUTION: Common throughout the state, except in the southern counties where it is found only occasionally.

There are some intermediate specimens between this variety and the following. In every case, however, where

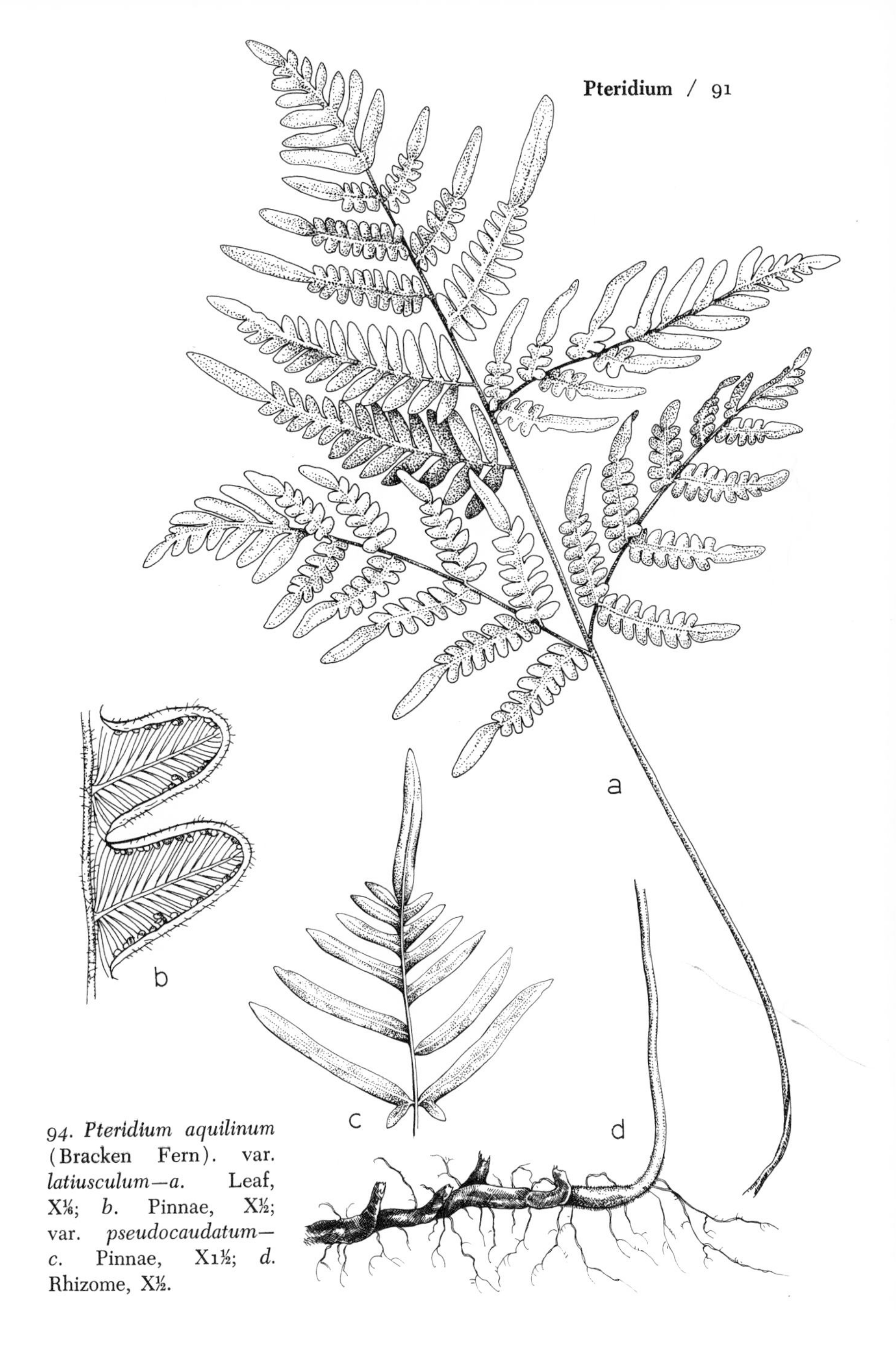

94. *Pteridium aquilinum* (Bracken Fern). var. *latiusculum—a.* Leaf, X⅙; *b.* Pinnae, X½; var. *pseudocaudatum—c.* Pinnae, X1½; *d.* Rhizome, X½.

the rhizome could be examined, there was no difficulty in determining to which taxon the specimen belonged.

1b. Pteridium aquilinum (L.) Kuhn var. **pseudocaudatum** (Clute) Heller, Cat. N. Am. Pl. 12. 1909. *Fig. 94c, d.*

Pteris aquilina var. *pseudocaudata* Clute, in Fern Bull. 8:39. 1900.

Growing tip of rhizome with a tuft of brown hairs; margin of ultimate leaf segments glabrous, to 4.5 mm broad.

COMMON NAME: Bracken Fern.
HABITAT: Open woodlands and fields.
RANGE: Massachusetts to Oklahoma, south to Texas and Florida.
ILLINOIS DISTRIBUTION: Rare; specimens definitely assignable to this taxon have been collected only from Hardin County (edge of oak woods, one mile north of Elizabethtown, August 14, 1963, *R. H. Mohlenbrock*).

4. *Cryptogramma* R. BR. – Cliffbrake Fern

Rhizome compact, slender; leaves bipinnate, somewhat dimorphic; sori marginal, continuous, covered by the reflexed margin of the leaf.

Only the following species occurs in Illinois.

1. Cryptogramma stelleri (S. G. Gmel.) Prantl, in Engler's Bot. Jahrb. 3:413. 1882. *Fig. 95.*

Pteris stelleri S. G. Gmel. in Nov. Comm. Acad. Sci. Petrop. 12:519. 1768.

Pteris gracilis Michx. Fl. Bor. Am. 2:262. 1803.

Allosorus gracilis (Michx.) Presl, Tent. Pterid. 153. 1836.

Pellaea gracilis (Michx.) Hook. Sp. Fil. 2:138. 1858.

Delicate perennial with creeping, slender, greenish rhizomes; blades bipinnate, thin, glabrous, the fertile to 20 cm long, the sterile slightly smaller and broader, the pinnules entire or lobed; petiole and rachis green, glabrous; sori continuous, covered by the reflexed leaf margin; spores greenish.

COMMON NAME: Slender Cliffbrake.
HABITAT: Deep, shaded, limestone ravines.
RANGE: Newfoundland to Alaska, south to Washington, Illinois, West Virginia, and New Jersey.
ILLINOIS DISTRIBUTION: Local; restricted to the northern one-fourth of Illinois.

This delicate fern, found primarily on deeply shaded limestone, has leaves of two types. The sterile leaves are shorter, broader, and more delicate in texture than the fertile ones.

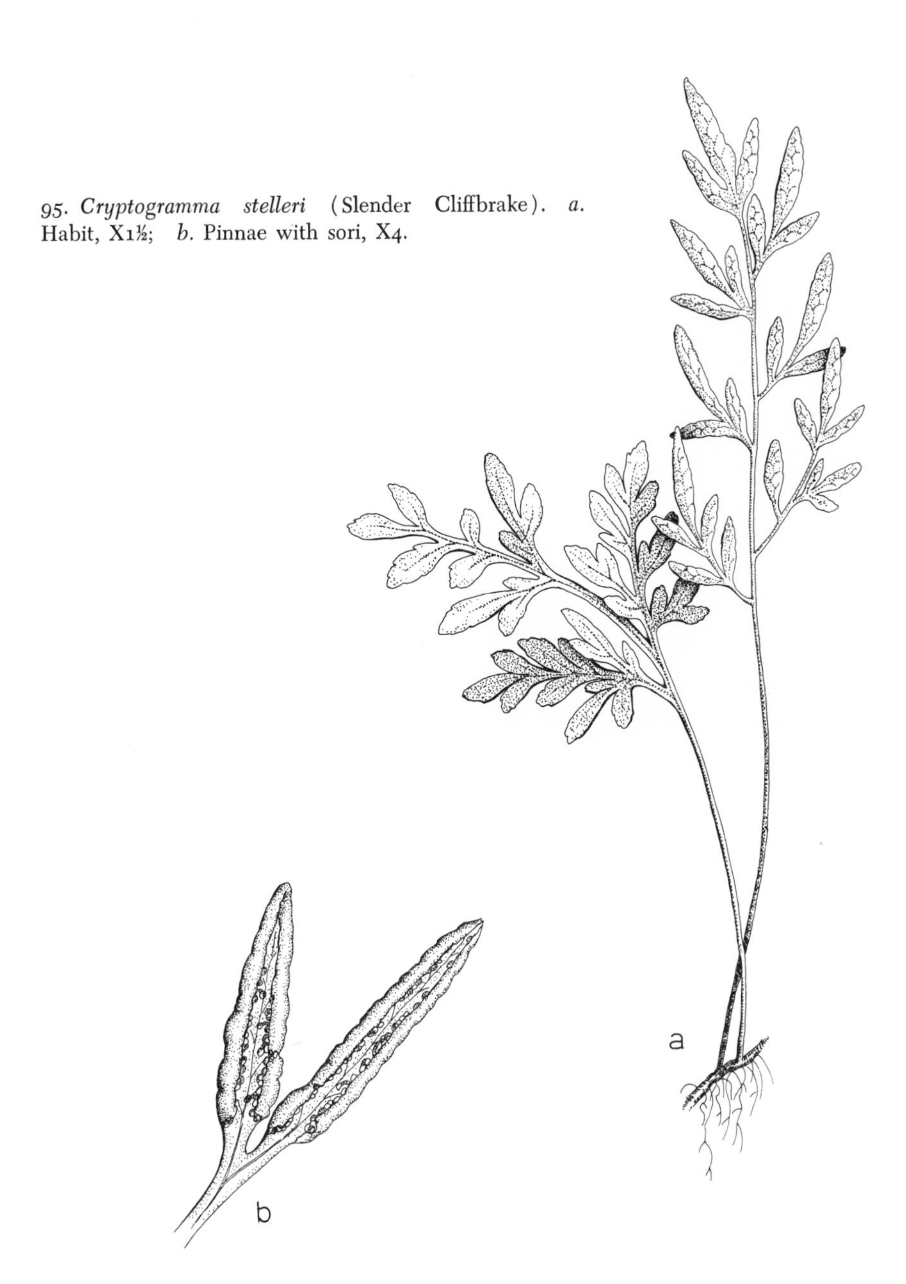

95. *Cryptogramma stelleri* (Slender Cliffbrake). *a.* Habit, X1½; *b.* Pinnae with sori, X4.

5. *Pellaea* LINK – Cliffbrake Fern

Rhizome compact, scaly; petiole and rachis purple-brown or dark brown; blade firm, once-pinnate to tripinnate; sori marginal, continuous, covered by the incurved leaf margin.

For a monographic study of *Pellaea* Section Pellaea, see Tryon (1957).

KEY TO THE SPECIES OF Pellaea IN ILLINOIS

1. Petiole and rachis pubescent nearly throughout; fertile pinnules much narrower than the sterile ones__________1. *P. atropurpurea*
1. Petiole and rachis glabrous or sparsely pubescent; fertile and sterile pinnae similar in shape____________________2. *P. glabella*

1. Pellaea atropurpurea (L.) Link, Fil. Sp. Hort. Berol. 59. 1841. *Fig. 96.*

Pteris atropurpurea L. Sp. Pl. 1076. 1753.

Allosorus atropurpureus (L.) Kunze ex Presl, Tent. Pterid. 153. 1836.

Rhizomes stocky; leaves rigid, erect, evergreen, pinnate to tripinnate, coriaceous, to 25 cm long; pinnules up to 12 pairs, entire or lobed, essentially glabrous; petiole and rachis pubescent, purple-brown, dull; sori marginal, continuous, covered (at least partially) by the incurved, irregular leaf margin; n = 87 (Britton, 1953).

COMMON NAME: Purple Cliffbrake.

HABITAT: On limestone cliffs, frequently under very dry conditions; rarely on sandstone (Ogle County), with much calcium deposition.

RANGE: Vermont to British Columbia, south to New Mexico and Florida; Mexico.

ILLINOIS DISTRIBUTION: In the counties bordering the Mississippi River in the southern half of Illinois; in the counties bordering the Ohio River in the southern one-fourth of Illinois; also Johnson and Ogle counties.

The incurved leaf margin, functioning as an indusium, is irregular in outline and frequently fails to cover all the sporangia. The pinnules are variable in degree of cutting. There is slight dimorphism with sterile pinnules tending to be shorter and broader. The petioles are always pubescent.

96. *Pellaea atropurpurea* (Purple Cliffbrake). *a*. Habit, X½; *b*. Portion of pinna with sori, X3.

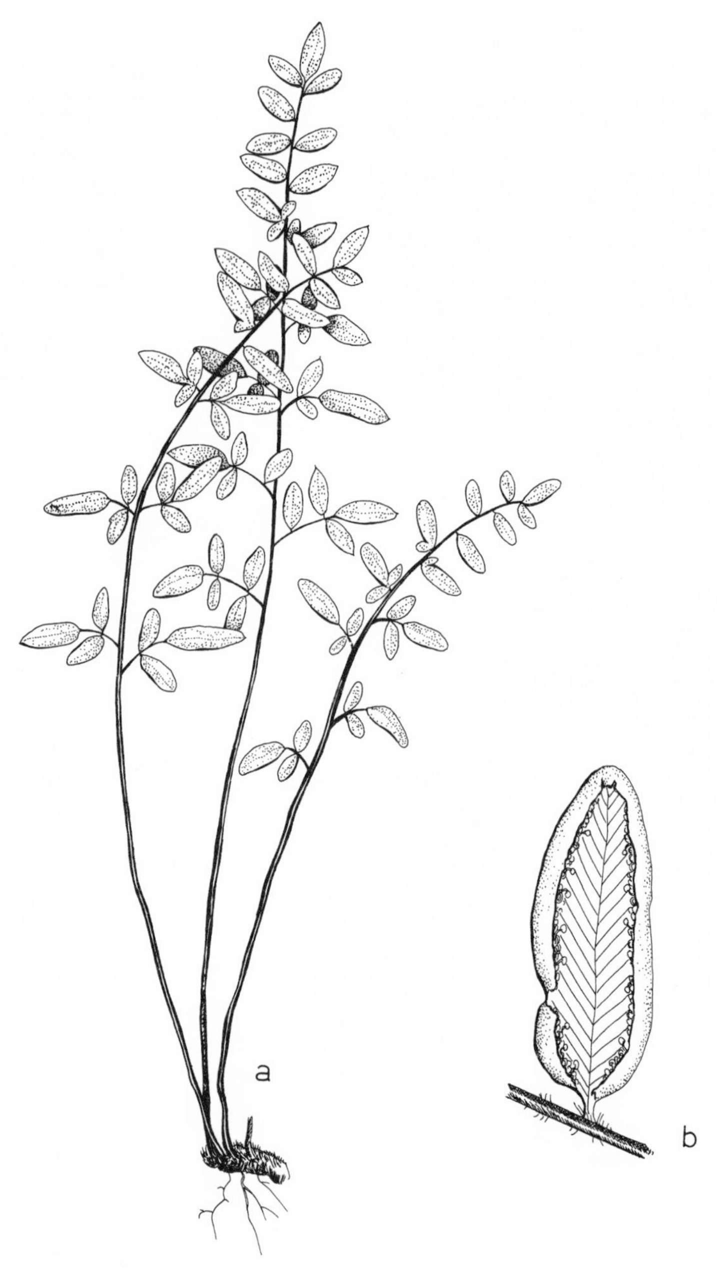

97. *Pellaea glabella* var. *glabella* (Smooth Cliffbrake). *a.* Habit, X½; *b.* Pinna with sori, X4.

2. Pellaea glabella Mett. ex Kuhn var. **glabella** *Fig.* 97.

Pellaea glabella Mett. ex Kuhn, in Linnaea 36:87. 1869.

Pellaea atropurpurea var. *bushii* Mack. ex Mack. & Bush, Man. Fl. Jackson Co. Mo. 5. 1902.

Pellaea atropurpurea f. *glabella* (Mett.) Clute, Our Ferns 386. 1938.

Leaves wiry, erect, evergreen, usually slightly smaller than in *P. atropurpurea,* pinnate or bipinnate, coriaceous, to 20 cm long; pinnae up to 10 pairs, usually lobed or auriculate, essentially glabrous; petiole and rachis sparsely hairy or glabrous, reddish-brown to dark brown, shining; sori marginal, continuous, covered by the incurved leaf margin; n = 116 (Britton, 1953).

COMMON NAME: Smooth Cliffbrake.

HABITAT: Limestone outcroppings.

RANGE: Ontario to South Dakota, south to Texas, Tennessee, and Virginia.

ILLINOIS DISTRIBUTION: Local in the northern counties, extending southward along the Mississippi and Ohio Rivers.

There does not seem to be any dimorphism of leaves in this species. It is sometimes rather difficult to distinguish this species from the usually larger and hairier *P. atropurpurea.*

6. *Cheilanthes* sw. – Lip Fern

Rhizomes short, compact, scaly; blades bipinnate-pinnatifid to tripinnate, hairy; sori marginal, more or less continuous, covered by the incurved leaf margin.

KEY TO THE SPECIES OF Cheilanthes IN ILLINOIS

1. Leaves densely brown-woolly beneath, to 15 cm long, tripinnate, with 7–12 (–15) pairs of pinnae; on limestone________1. *C. feei*
1. Leaves white-villous beneath, some usually well over 15 cm long, bipinnate-pinnatifid, with 12–20 pairs of pinnae; chiefly on sandstone____2. *C. lanosa*

1. Cheilanthes feei Moore, Ind. Fil. 38. 1857. *Fig.* 98.

Cheilanthes lanuginosa Nutt. ex Hook. Sp. Fil. 2:99. 1858.

Small, tufted evergreen perennial from short rhizomes; leaves bipinnate-pinnatifid to tripinnate, to 15 cm long; blades densely

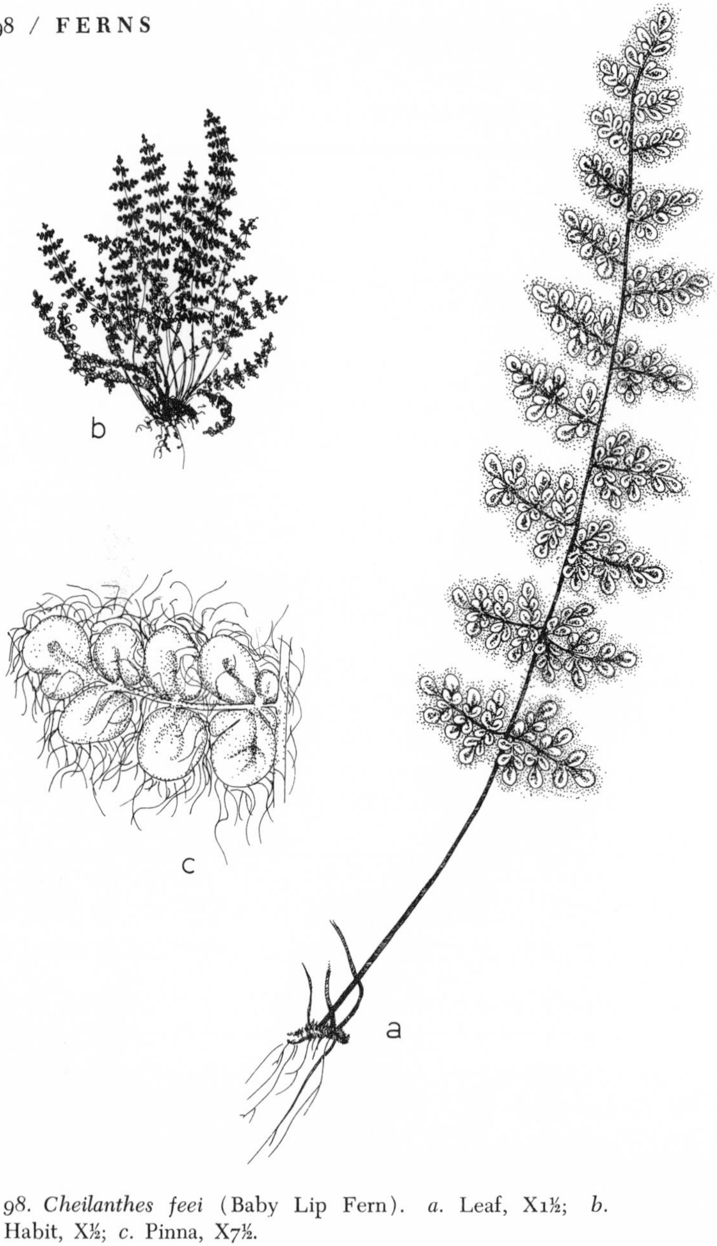

b

c

a

98. *Cheilanthes feei* (Baby Lip Fern). *a.* Leaf, X1½; *b.* Habit, X½; *c.* Pinna, X7½.

woolly, broadest near base, with 7–15 pairs of pinnae, each with 2–5 pairs of pinnules; petiole and rachis brown, pubescent at first, becoming nearly glabrous; sori marginal, covered by the incurved leaf margin.

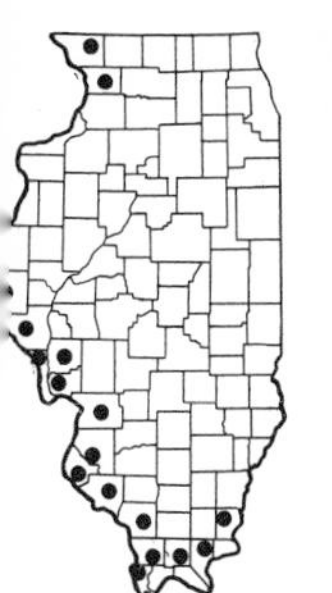

COMMON NAME: Baby Lip Fern; Slender Lip Fern; Resurrection Fern.

HABITAT: Dry, exposed limestone cliffs.

RANGE: Wisconsin to British Columbia, south to California, Texas, and Arkansas.

ILLINOIS DISTRIBUTION: In counties bordering the Mississippi River in the southern half of Illinois; in a few counties bordering the Ohio River; also JoDaviess and Carroll counties.

This is one of the smallest members of Polypodiaceae in Illinois. During extreme dry conditions, this fern has the ability to curl up into a tiny, dry mat, only to expand when additional moisture becomes available.

2. **Cheilanthes lanosa** (Michx.) D. C. Eaton in Torr. in Rep. U. S. & Mex. Bound. Surv. 2:234. 1859. *Fig.* 99.

Nephrodium lanosum Michx. Fl. Bor. Am. 2:270. 1803, pro parte.

Adiantum vestitum Spreng. in Anleit. Kennt. Gewächse 3:122. 1804, pro parte.

Cheilanthes vestita (Spreng.) Sw. Syn. Fil. 128. 1806.

Tufted evergreen perennial from short-creeping rhizomes; leaves bipinnate-pinnatifid, to 30 cm long; blades sparsely hairy above, densely white-villous beneath, with up to 20 pairs of pinnae, each with 5–9 pairs of pinnules; petiole and rachis brown, usually somewhat pubescent; sori marginal, covered by the incurved leaf margin.

COMMON NAME: Hairy Lip Fern.

HABITAT: Dry, exposed cliffs, usually sandstone, rarely limestone.

RANGE: Connecticut to Minnesota, south to Texas and Georgia.

ILLINOIS DISTRIBUTION: Local; restricted to the southern one-fourth of the state.

This species is hairy, but not woolly like *C. feei.* Fernald (1950) has created confusion by erroneously calling this species *C. vestita,* and applying the binomial *C. lanosa* to a more southern species.

The usual habitat is on exposed sandstone, but a few stations are known for this species which are calcareous.

99. *Cheilanthes lanosa* (Hairy Lip Fern). *a.* Leaf, X1; *b.* Habit, X¼; *c.* Pinna, X7.

7. *Polypodium* [Tourn.] L. – Polypody

Rhizomes cord-like, scaly and knobby; blades pinnatifid, glabrous or scaly; sori round, separate, borne on the back of the leaf; indusium none.

KEY TO THE SPECIES OF Polypodium IN ILLINOIS

1. Blades and petioles scaleless, appearing smooth, the segments minutely toothed________________________________1. *P. vulgare*
1. Lower surface of blades and petioles scaly, appearing pustular, the segments entire___________________________2. *P. polypodioides*

1. **Polypodium vulgare** L. var. **virginianum** (L.) Eaton in Gray, Man. 373. 1818. *Fig. 100.*

Polypodium virginianum L. Sp. Pl. 1085. 1753.
Polypodium vulgare var. *americanum* Hook. Fl. Bor. Am. 2:258. 1840.

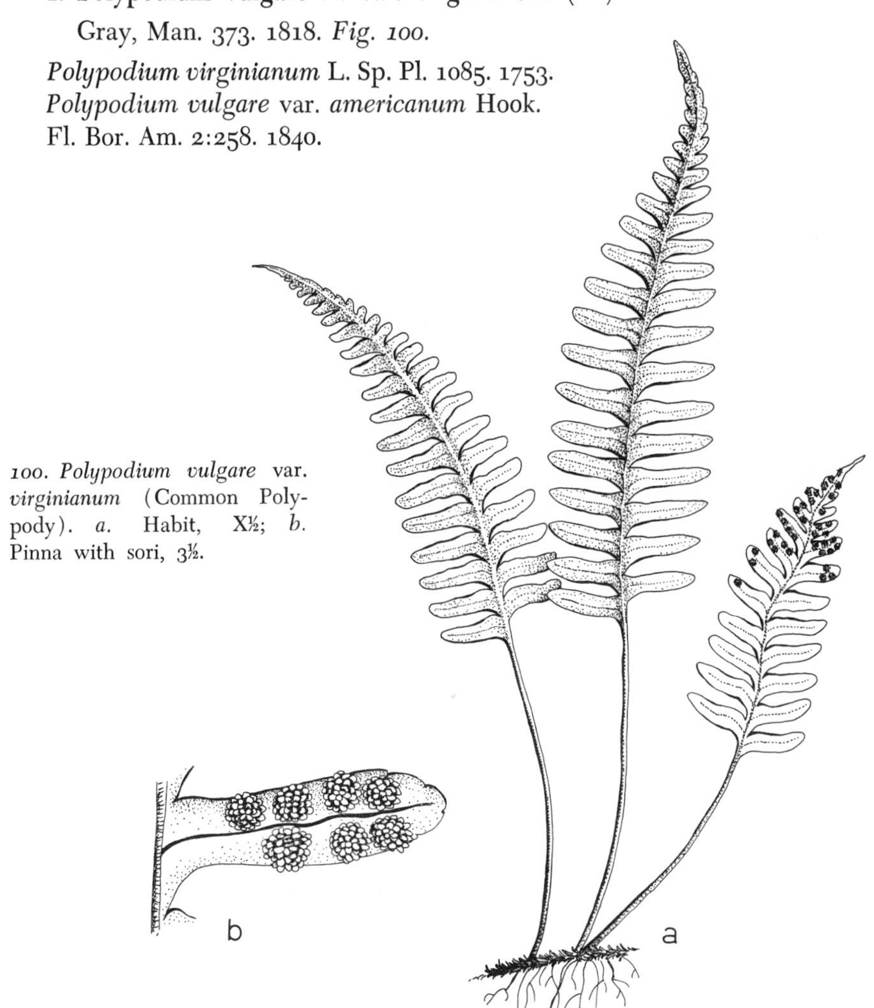

100. *Polypodium vulgare* var. *virginianum* (Common Polypody). *a.* Habit, X½; *b.* Pinna with sori, 3½.

Evergreen from rather spongy, scaly rhizomes; leaves pinnatifid, to 25 cm long; segments lanceolate to lance-oblong, obtuse or acute, more or less entire, glabrous; petiole slender, glabrous; sori borne on the back of the pinnae near the margin, round; indusium none; n = 37 (Manton, 1950).

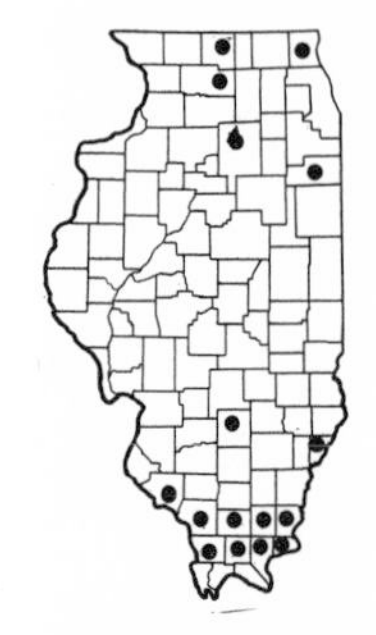

COMMON NAME: Common Polypody.
HABITAT: Sandstone rock, usually under some shade.
RANGE: Newfoundland to British Columbia, south to Arkansas and Georgia.
ILLINOIS DISTRIBUTION: Occasional in the southern one-third of Illinois; local and rare in the northern one-third; absent from the central one-third.

This species and the following are found commonly on sandstone rock, often forming rather extensive colonies. Considerable variation exists in shape and margin of pinnae. Many of the extreme forms have been given names, but none of these apparently is in Illinois. Variety *virginianum* may be used to distinguish our plants from European ones.

2. **Polypodium polypodioides** (L.) Watt var. **michauxianum** Weatherby, in Contr. Gray Herb. 124:31. 1939. *Fig. 101.*

Polypodium incanum Sw. Prodr. 131. 1788.

Polypodium ceteraccinum Michx. Fl. Bor. Am. 2:271. 1803.

Evergreen from slender, scaly rhizomes; leaves pinnatifid, to 20 cm long; pinnae narrowly oblong, obtuse, entire, glabrous above, gray-scaly beneath; petiole slender, scurfy; sori borne on the back of the leaf, round; indusium none.

COMMON NAME: Gray Polypody; Resurrection Fern.
HABITAT: Sandstone rocks; tree trunks and branches.
RANGE: Delaware to Oklahoma, south to Texas and Florida; Central America.
ILLINOIS DISTRIBUTION: Not common; restricted to the southern one-fourth of the state.

The scurfy-pustular appearance of the blades readily distinguishes this species, which is somewhat smaller than the preceding. In addition to its rock habitat, this species may grow epiphytically on the trunks and branches of trees. During drought conditions, the leaves curl, only to revive when moisture becomes available.

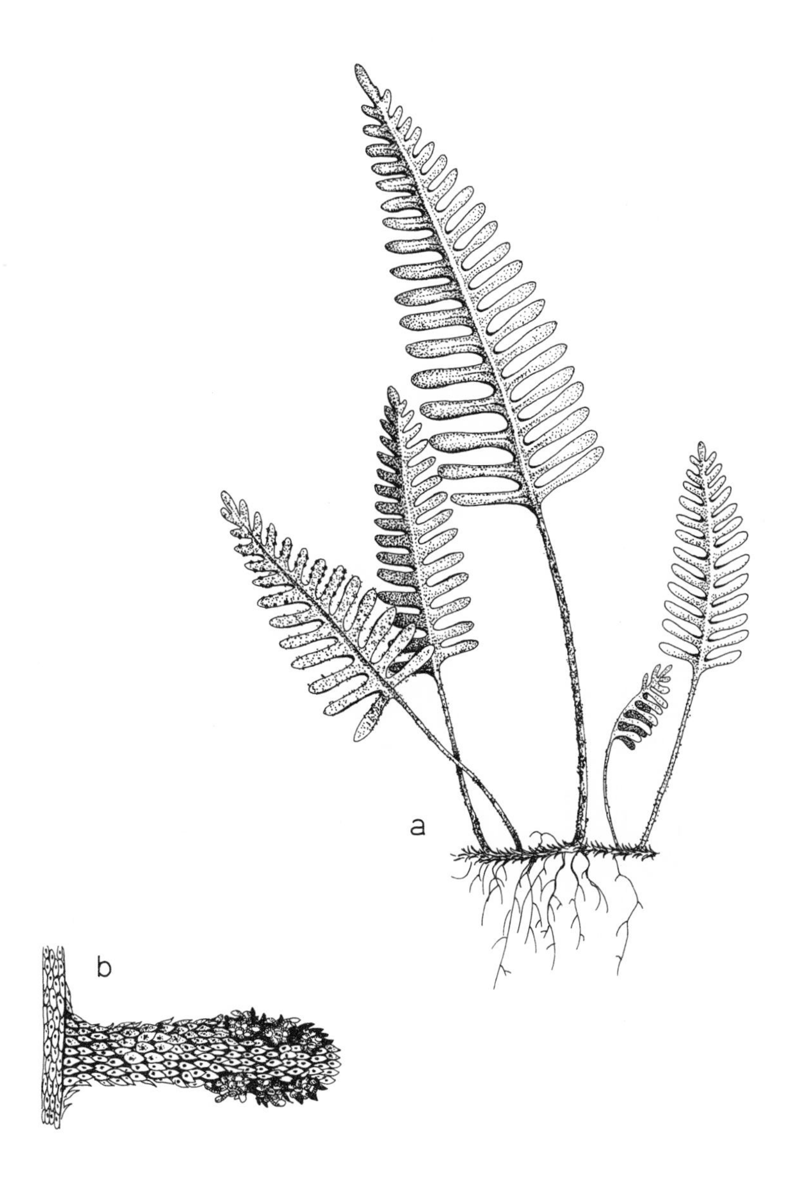

101. *Polypodium polypodioides* var. *michauxianum* (Gray Polypody). *a*. Habit, X1; *b*. Pinna with scales and sori, X3¼.

8. *Polystichum* ROTH – Christmas Fern

Rhizome stout, thick, densely scaly; blades once-pinnate; sori round, confined to the upper, reduced pinnae; indusium attached centrally in each sorus.

Only the following species occurs in Illinois.

1. **Polystichum acrostichoides** (Michx.) Schott, Gen. Fil. pl. 9. 1834. *Fig. 102.*

Nephrodium acrostichoides Michx. Fl. Bor. Am. 2:267. 1803.
Aspidium acrostichoides (Michx.) Sw. Syn. Fil. 44. 1806.
Aspidium schweinitzii Beck, Bot. U. S. 449. 1833.
Polystichum acrostichoides var. *incisum* Gray, Man. 632. 1848.
Polystichum acrostichoides var. *schweinitzii* (Beck) Small, in Bull. Torrey Club 20:464. 1893.
Polystichum acrostichoides f. *incisum* (Gray) Gilb. List N. Am. Pterid. 19. 1901.

Stout evergreen perennial from thick, scaly rhizomes covered by the bases of old petioles; leaves once-pinnate, to 50 cm long, lanceolate, with up to over 30 pairs of pinnae; pinnae lanceolate to oblong, obtuse to acute, glabrous and shiny above, short-petiolulate, minutely or coarsely serrate, the teeth bristle-tipped, with a basal auricle, the lowermost pinnae sterile and larger than the upper fertile ones; sori distinct or confluent, with the indusium centrally attached in each sorus.

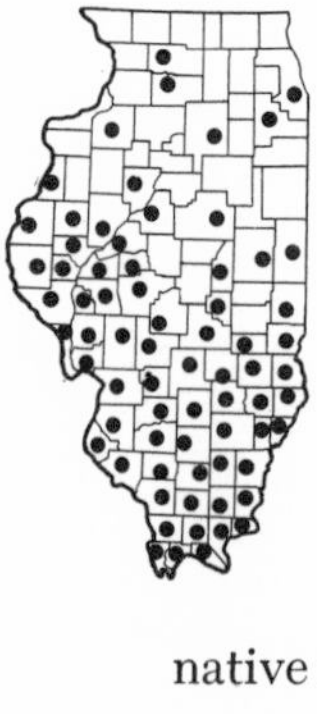

COMMON NAME: Christmas Fern.
HABITAT: Woodlands, particularly in rocky soil.
RANGE: Prince Edward Island to Wisconsin, south to Texas and Florida.
ILLINOIS DISTRIBUTION: Common; known from every county, except in the northern one-third of the state.

Plants which have the blades coarsely serrate may be distinguished as f. *incisum*. This taxon occurs occasionally throughout Illinois.

Christmas fern is one of the most familiar and popular native ferns. The common name is derived from the fact that pioneers would gather the evergreen fronds at Christmas for decorative purposes.

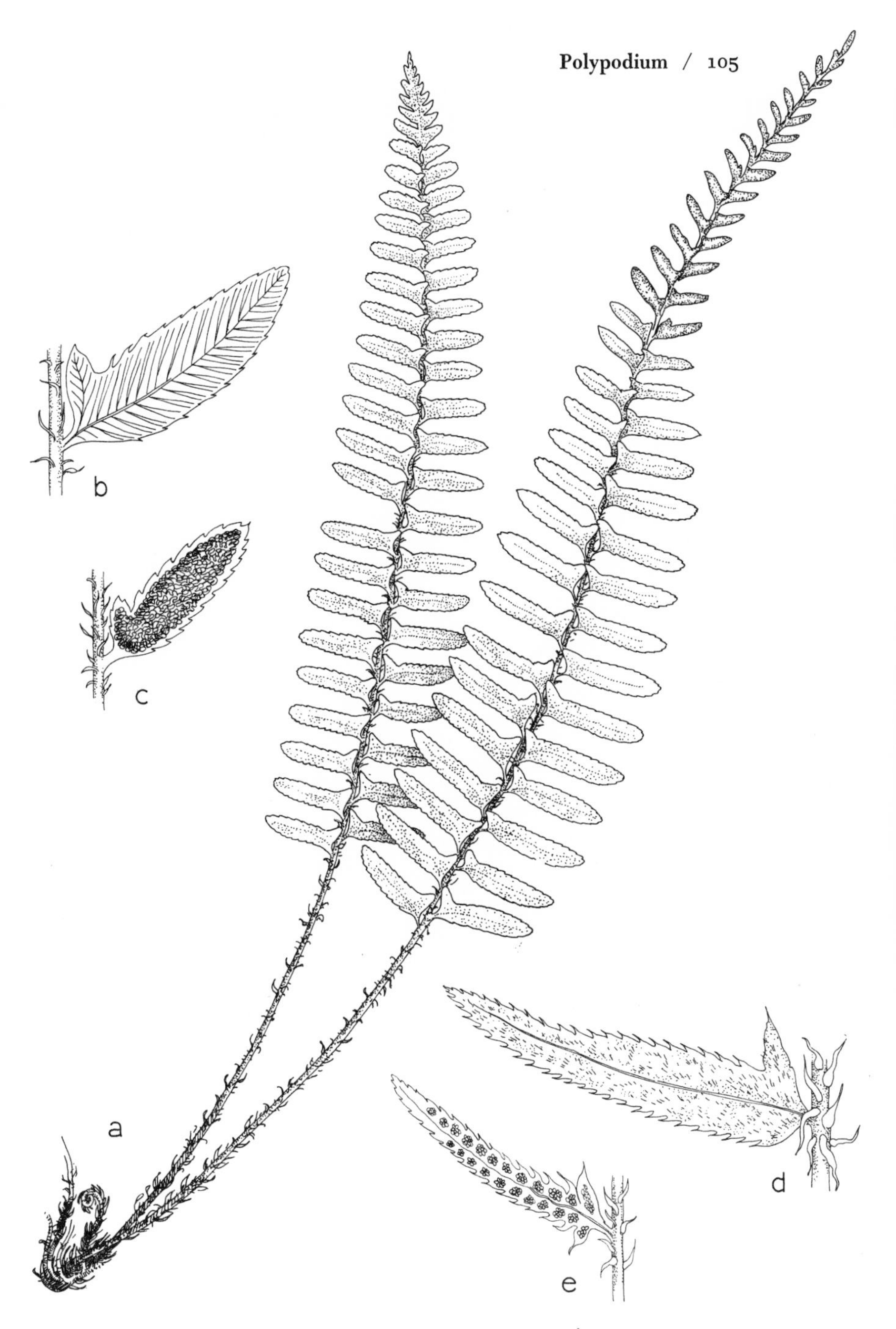

102. *Polystichum acrostichoides* (Christmas Fern). *a.* Habit, X¼; *b.* Sterile pinna, X1¼; *c.* Fertile pinna, X2; forma *incisum*—*d.* Sterile pinna, X1½; *e.* Fertile pinna, X1½.

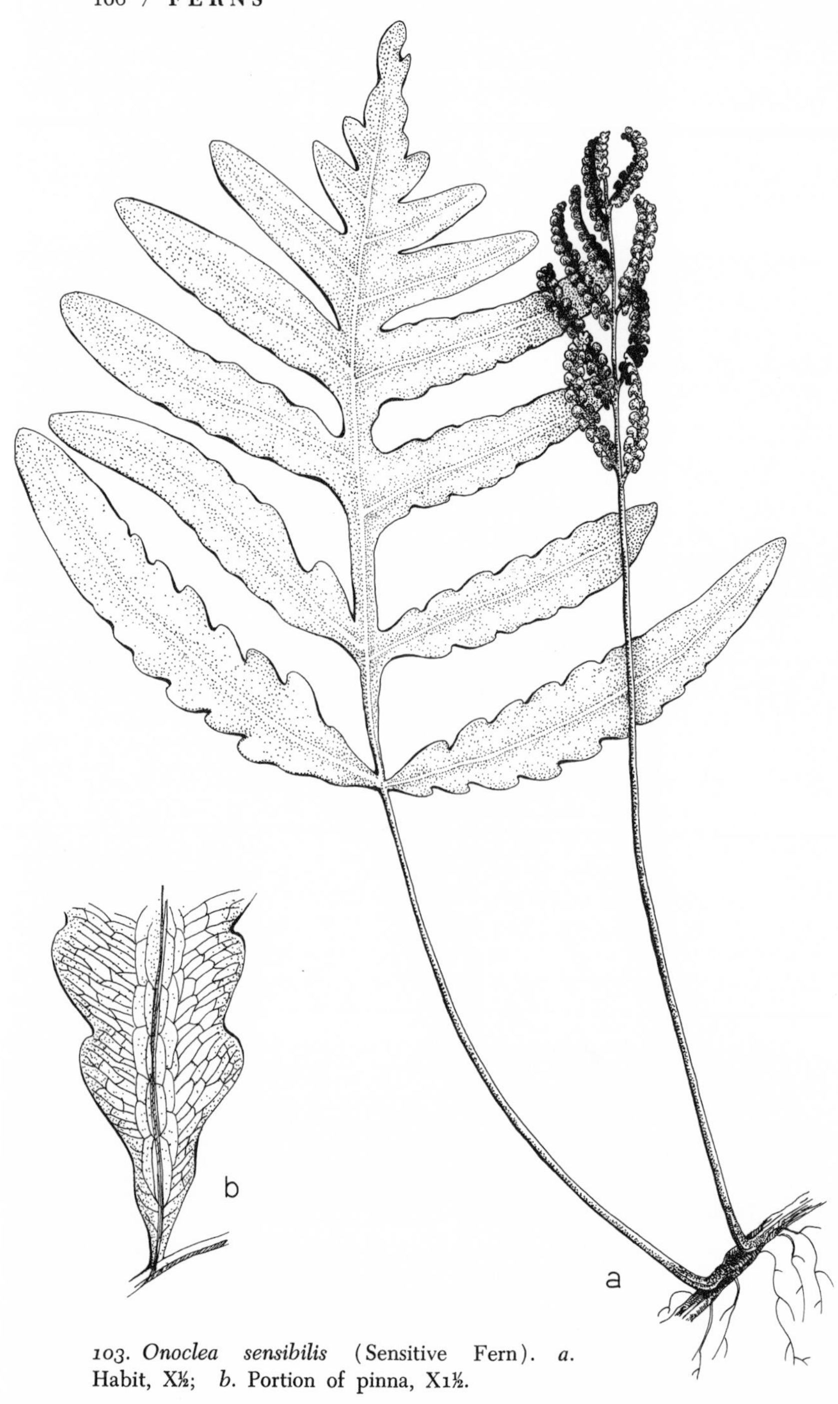

103. *Onoclea sensibilis* (Sensitive Fern). *a.* Habit, X½; *b.* Portion of pinna, X1½.

9. *Onoclea* L. – Sensitive Fern

Rhizomes branched and creeping, scaleless; leaves dimorphic, the sterile pinnatifid, the fertile bipinnate; sori round, covered by a fragile indusium, the sporangia becoming hard at maturity.

Only the following species occurs in Illinois.

1. **Onoclea sensibilis** L. Sp. Pl. 1062. 1753. *Fig. 103.*

Deciduous perennial from creeping, branching rhizomes; leaves dimorphic, the sterile pinnatifid, to 50 cm long, with a winged rachis, the fertile bipinnate, the stiff pinnules tightly enrolled to enclose the sporangia; sori round, covered by a delicate, hood-like indusium; fertile pinnules at maturity dry, hard, globoid, eventually cracking to liberate the spores.

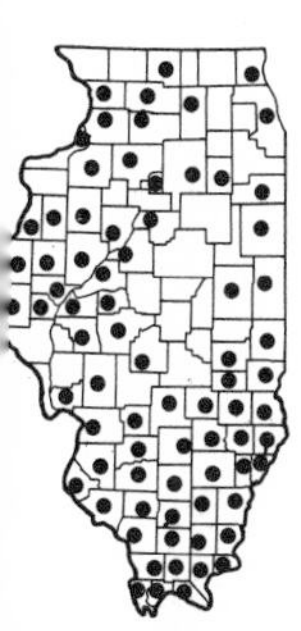

COMMON NAME: Sensitive Fern.
HABITAT: Moist woodlands or low, open ground.
RANGE: Newfoundland to Manitoba, south to Texas and Florida.
ILLINOIS DISTRIBUTION: Common; in nearly every county.

The dimorphic nature of this fern readily permits identification. The veins of the sterile blades form a network, rather than remaining free as in most members of the Polypodiaceae. The sterile leaves are deciduous (hence the common name), while the fertile leaves persist through the winter.

10. *Matteuccia* TODARO – Ostrich Fern

Rhizome stout, branched and creeping, scaly; leaves dimorphic, the outer large and sterile; sporangia globoid, borne on the revolute margins of the pinnules.

Only the following species occurs in Illinois.

1. **Matteuccia struthiopteris** (L.) Todaro, in Giorn. Sci. Nat. Palermo 1:235. 1866. *Fig. 104.*

Osmunda struthiopteris L. Sp. Pl. 1066. 1753.
Struthiopteris filicastrum All. Fl. Ped. 2:283. 1785.
Onoclea struthiopteris (L.) Hoffm. Deutsch. Fl. 2:11. 1795.
Onoclea nodulosa Michx. Fl. Bor. Am. 2:272. 1803.
Struthiopteris germanica Willd. Enum. Pl. Hort. Berol. 1071. 1809.
Struthiopteris pensylvanica Willd. Sp. Pl. 5:289. 1810.

Pteretis struthiopteris (L.) Nieuwland, in Am. Midl. Nat. 3:197. 1914.
Matteuccia nodulosa (Michx.) Fern. in Rhodora 17:164. 1915.
Pteretis nodulosa (Michx.) Nieuwland, in Am. Midl. Nat. 4:334. 1916.
Pteretis pensylvanica (Willd.) Fern. in Rhodora 47:123. 1945.
Matteuccia pensylvanica (Willd.) Raymond, in Nat. Can. 77:55. 1950.
Matteuccia struthiopteris var. *pensylvanica* (Willd.) Morton, in Am. Fern Journ. 40:247. 1950.

Robust, handsome, evergreen perennial from stout, creeping and branching, scaly rhizomes; leaves dimorphic, the sterile ones forming a circle around the fertile ones, to nearly 2 m tall, pinnate-pinnatifid, the pinnae acuminate, glabrous, the petioles green, glabrous, 4-angled; fertile leaves stiffly erect, to 50 cm tall, dark green to blackish, with twisted, bead-like pinnae surrounding the sporangia, the petioles brown and shining.

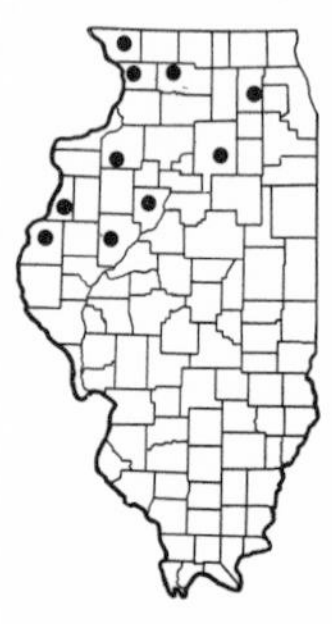

COMMON NAME: Ostrich Fern.
HABITAT: Rich, moist woodlands.
RANGE: Newfoundland to Alaska, south to British Columbia, Missouri, and Virginia.
ILLINOIS DISTRIBUTION: Rare; restricted to the northern one-third of the state, usually in the western counties.

This is the largest and most handsome fern in Illinois. Individual plants form magnificent, huge, spreading foliage.

The nomenclature has been confused repeatedly with the result that this plant has been known by many names. Because of the similar features of their fertile pinnae, this species and *Onoclea sensibilis* probably are rather closely related.

11. *Gymnocarpium* NEWM. – Oak Fern

Rhizome slender, creeping; leaves bipinnate-pinnatifid; sori round, rather small; indusium absent.

The lack of an indusium, combined with the absence of wings on the glabrous rachis, separate this genus from *Thelypteris* and *Dryopteris*.

The inability to define consistently this and the following two genera has created much nomenclatural confusion.

Only the following species occurs in Illinois.

104. *Matteuccia struthiopteris* (Ostrich Fern). *a.* Habit, X⅛; *b.* Pinnae, 1½; *c.* Fertile leaf, X⅙; *d.* Fertile pinna, X¾.

105. *Gymnocarpium dryopteris* (Oak Fern). *a.* Leaf, X⅔; *b.* Pinna with sori, X5.

1. **Gymnocarpium dryopteris** (L.) Newm. in Phytologist 4: app. 24. 1851. *Fig. 105.*

Polypodium dryopteris L. Sp. Pl. 1093. 1753.

Nephrodium dryopteris (L.) Michx. Fl. Bor. Am. 2:270. 1803.

Phegopteris dryopteris (L.) Fee, Gen. Fil. 243. 1850–52.

Polypodium dryopteris var. *disjunctum* Ledeb. Fl. Rosc. 4:509. 1853.

Polypodium disjunctum (Ledeb.) Rupr. ex Schur, in Oesterr. Bot. Zeit. 8:193. 1858.

Dryopteris disjuncta (Ledeb.) Morton, in Rhodora 43:217. 1941.

Deciduous perennial from slender, scaly rhizomes; leaves deltoid, ternate, the lateral divisions divergent and with longer pinnae on the lower side; lowest pinnae of each division the largest, membranous, linear-oblong, obtuse, glabrous, sessile; petiole and rachis very slender, glabrous, except for the sparsely chaffy petiole-base; sori small, borne on the lower surface of the pinnae near the margin; indusium none; n = 80 (Manton, 1950), 40 (Wagner & Chen, 1964).

COMMON NAME: Oak Fern.

HABITAT: Deep, moist woodlands.

RANGE: Greenland to Alaska, south to Oregon, Arizona, Missouri, and Virginia.

ILLINOIS DISTRIBUTION: Very rare and probably extinct; St. Clair and Ogle counties (rocky woods, near Oregon, August 11, 1885, *M. B. Waite*), not seen since 1885.

Taxonomists do not agree on the proper generic disposition of this species, with some placing it in *Dryopteris*, some in *Thelypteris*, some in *Currania*, and others in *Gymnocarpium*.

12. *Thelypteris* SCHMIDEL – Beech Fern

Rootstock slender, creeping or erect, branched; leaves pinnate-pinnatifid to bipinnate-bipinnatifid, sometimes triangular; rachis winged or unwinged; sori round; indusium none or cordate.

This genus is often combined with *Dryopteris*. It differs from *Dryopteris* in the deciduous nature of the leaves. An indusium is often wanting.

KEY TO THE SPECIES OF *Thelypteris* IN ILLINOIS

1. Rachis winged throughout, or only the basal pair of pinnae stalked; indusium none.
   2. Rachis winged only above the two basal pinna pairs, the wings not extending to the lowest pair of pinnae; blades pinnate-pinnatifid; rachis rather densely brown-scaly__1. *T. phegopteris*
   2. Rachis winged throughout; blades bipinnatifid; rachis sparsely white-scaly____________________________2. *T. hexagonoptera*
1. Rachis not winged, the pinnae separate from it nearly to apex of blade; indusium cordate.
   3. Lowest pinnae strongly reduced________3. *T. noveboracensis*
   3. Lowest pinnae only slightly reduced or not at all____4. *T. palustris*

1. Thelypteris phegopteris (L.) Slosson ex Rydb. Fl. Rocky Mts. 1043. 1917. *Fig. 106.*

Polypodium phegopteris L. Sp. Pl. 1089. 1753.
Polypodium connectile Michx. Fl. Bor. Am. 2:271. 1803.
Phegopteris polypodioides Fee, Gen. Fil. 243. 1850–52.
Gymnocarpium phegopteris (L.) Newm. in Phytologist 4:app. 23. 1851.
Phegopteris connectilis (Michx.) Watt, in Can. Nat. II. 3:159. 1866.
Dryopteris phegopteris (L.) C. Chr. Ind. Fil. 284. 1905.

Rather delicate, deciduous perennial from slender, branching rhizomes; leaves triangular, tapering to the tip, pinnate-pinnatifid, to 45 cm long, the pinnae up to 25 pairs, pubescent beneath (also occasionally above), connected by a winged rachis except for the lowest pair of pinnae; petiole slender, sparsely or rather densely brown-scaly to glabrous; sori small, round, borne on the back of the leaf segments near the margin; indusium absent.

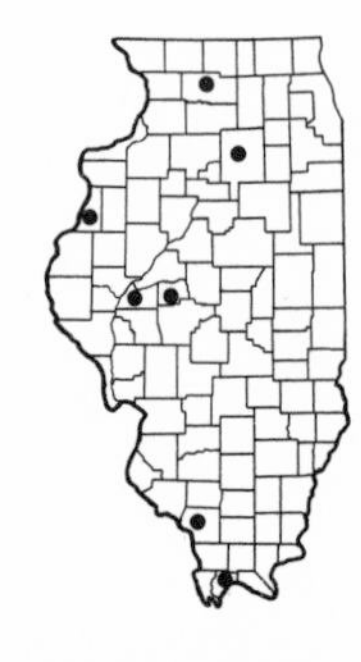

COMMON NAME: Long Beech Fern.
HABITAT: Moist, shaded woodlands in sandstone areas.
RANGE: Greenland to Alaska, south to Oregon, Illinois, and North Carolina.
ILLINOIS DISTRIBUTION: Local; in seven widely scattered counties.
The absence of an indusium and the triangular fronds are characters which relate this fern very closely to

T. hexagonoptera. The latter, however, has the winged rachis extending all the way to the lowest pair of pinnae.

The absence of an indusium and the triangular fronds has caused some botanists to place these species in *Phegopteris.*

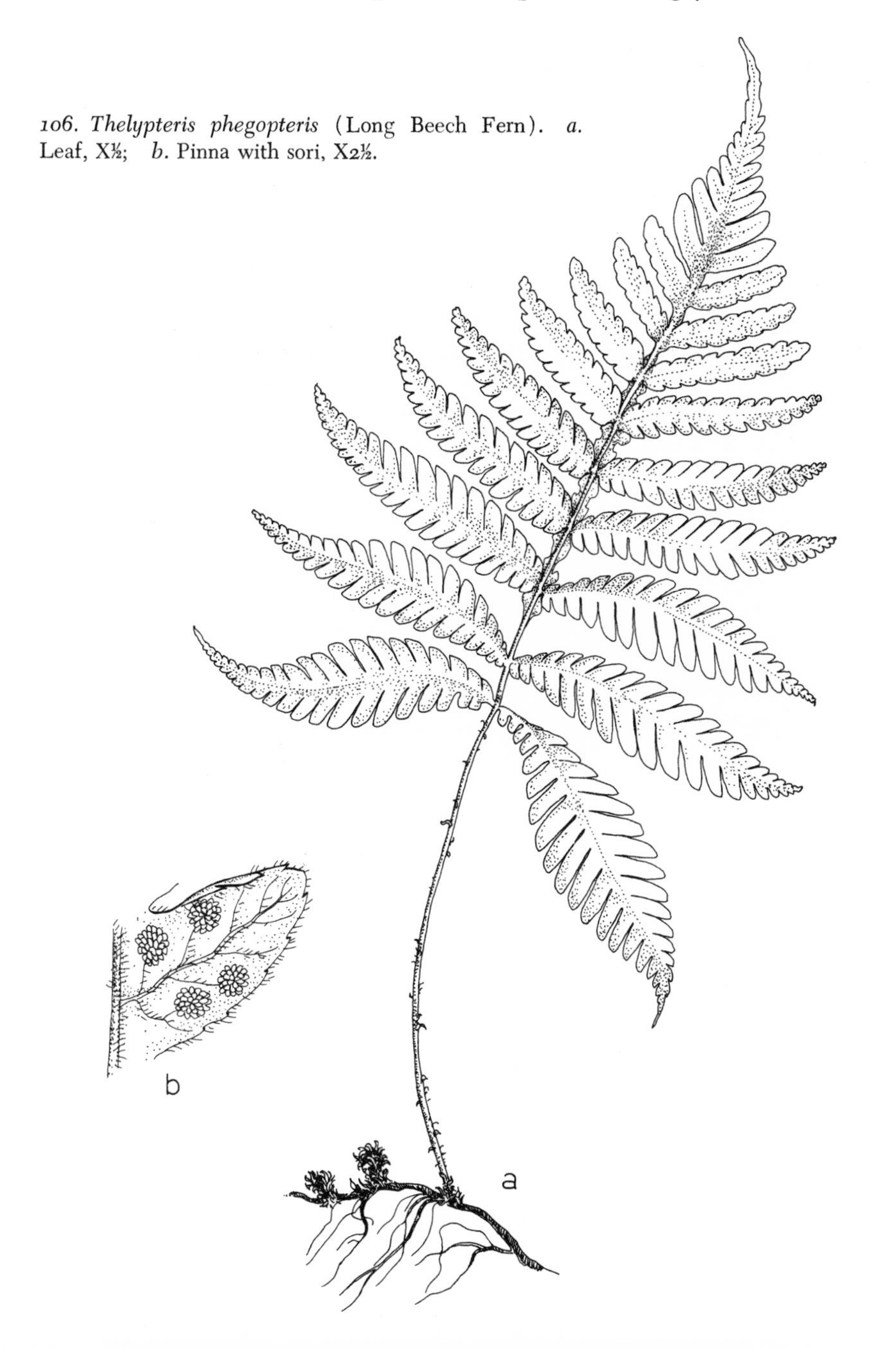

106. *Thelypteris phegopteris* (Long Beech Fern). *a.* Leaf, X½; *b.* Pinna with sori, X2½.

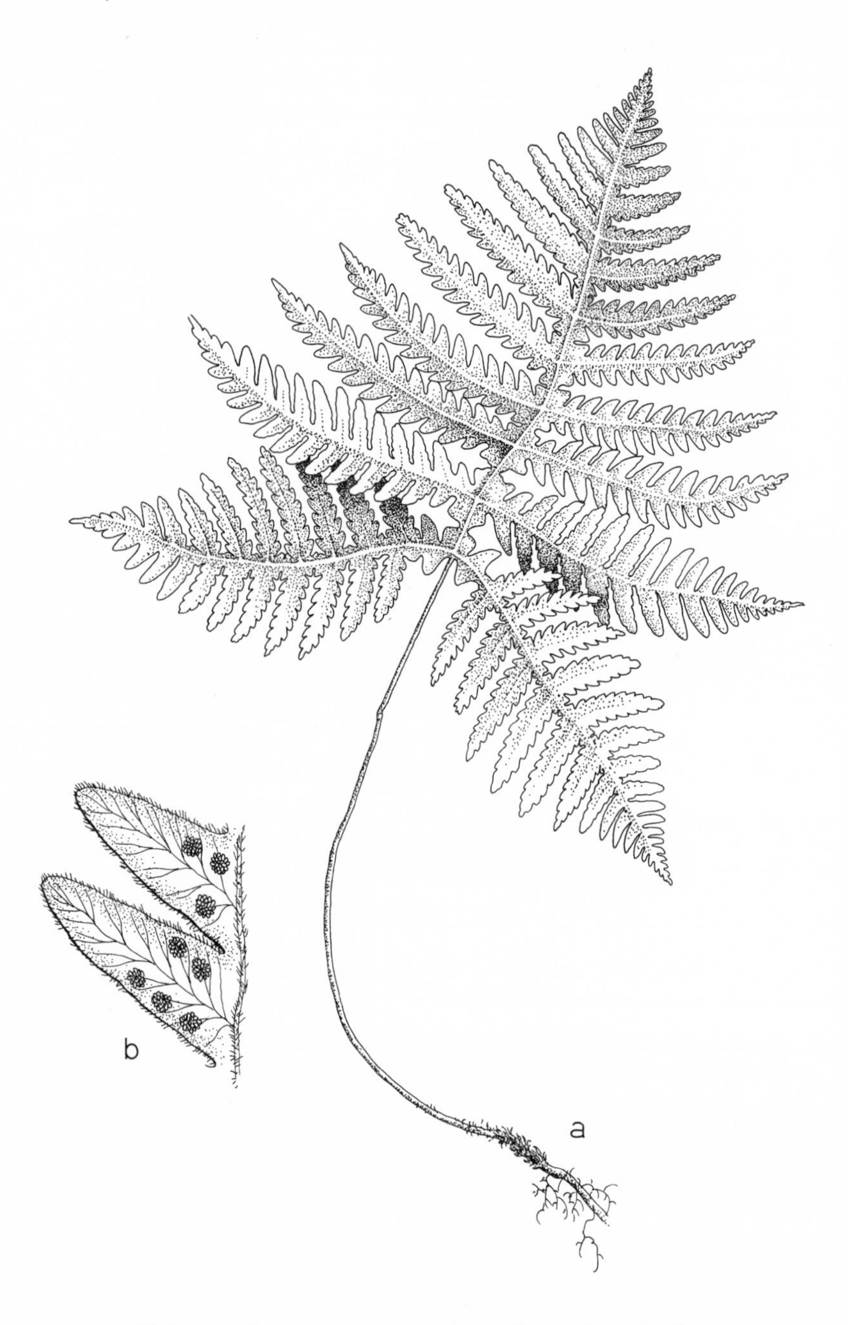

107. *Thelypteris hexagonoptera* (Broad Beech Fern). *a.* Leaf, X½; *b.* Pinnae with sori, X2½.

2. **Thelypteris hexagonoptera** (Michx.) Weatherby, in Rhodora 21:179. 1919. *Fig. 107.*

Polypodium hexagonopterum Michx. Fl. Bor. Am. 2:271. 1803.
Phegopteris hexagonoptera (Michx.) Fee, Gen. Fil. 243. 1850–52.
Dryopteris hexagonoptera (Michx.) C. Chr. Ind. Fil. 270. 1905.

Rather delicate, deciduous perennial from branching rhizomes; leaves triangular, tapering to the tip, bipinnatifid, to 50 cm long, the pinnae up to 25 pairs, glandular-puberulent beneath, connected by a winged rachis which extends to the lowest pair of pinnae; petiole slender, sparsely scaly to glabrous; sori small, round, borne on the back of leaf segments; indusium absent.

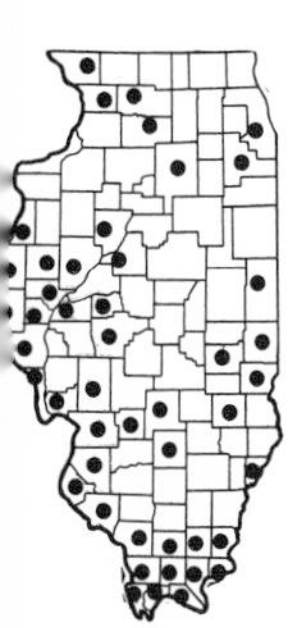

COMMON NAME: Broad Beech Fern.
HABITAT: Rich woodlands.
RANGE: Quebec to Ontario, south to Texas and Florida.
ILLINOIS DISTRIBUTION: Fairly common throughout the state, but seemingly absent from many of the east-central counties.

The completely winged rachis separates this species from *T. phegopteris.*

This species has been used by the author for years as an indicator plant for various native orchids in Illinois. Dr. Wagner has reported that he uses this species as an indicator for *Athyrium pycnocarpon, A. thelypterioides,* and *Dryopteris goldiana.*

3. **Thelypteris noveboracensis** (L.) Nieuwland, in Am. Midl. Nat. 1:226. 1910. *Fig. 108.*

Polypodium noveboracense L. Sp. Pl. 1091. 1753.
Aspidium noveboracense (L.) Sw. in Journ. Bot. Schrad. 1800 (2):38. 1801.
Nephrodium noveboracense (L.) Desv. in Mem. Soc. Linn. 6:257. 1827.
Dryopteris noveboracensis (L.) Gray, Man. 630. 1848.

Delicate, deciduous perennial from slender rhizomes; leaves tapering to both ends, pinnate-pinnatifid, to 50 cm long, with up to 40 pairs of pinnae, the pinnae membranous, puberulent beneath, the lowermost pinnae strongly reduced; petiole slender, pale, glabrous; sori distinct, borne on the back of the leaf segments; indusium glandular.

b

a

108. *Thelypteris noveboracensis* (New York Fern). *a.* Leaf, X⅓; *b.* Pinna with sori, X6.

COMMON NAME: New York Fern.
HABITAT: Moist, rarely dry, woodlands.
RANGE: Newfoundland to Ontario, south to Arkansas and Georgia.
ILLINOIS DISTRIBUTION: Rare; known only from two northeastern counties and three southern counties.
The much reduced basal pinnae well define this handsome, fragile fern.

4. **Thelypteris palustris** Schott var. **pubescens** (Laws.) Fern. in Rhodora 31:34. 1929. *Fig. 109.*

Acrostichum thelypteris L. Sp. Pl. 1071. 1753.
Aspidium thelypteris (L.) Sw. in Journ. Bot. Schrad. 1800 (2):40. 1801.
Nephrodium thelypteris (L.) Strempel, Fil. Berol. Syn. 32. 1822.
Dryopteris thelypteris (L.) Gray, Man. 630. 1848.
Lastrea thelypteris α pubescens Laws. in Edinb. New Philos. Journ. II. 19:277. 1864.
Dryopteris thelypteris var. *pubescens* (Laws.) A. R. Prince ex Weatherby, in Am. Fern Journ. 26:95. 1936.

Rather fragile, deciduous perennial from slender, nearly scaleless rhizomes; leaves somewhat dimorphic, the sterile pinnate-pinnatifid, membranous, to 50 cm long, puberulent on both surfaces, with up to 40 pairs of pinnae; fertile leaves pinnate-pinnatifid, more firm, to 40 cm long, puberulent on both surfaces, with up to 25 pairs of pinnae; petiole slender, glabrous at maturity; sori round, usually confluent, borne on the back of the leaf segments; indusium eglandular; n = 35.

COMMON NAME: Marsh Fern.
HABITAT: Marshy ground and in swamps.
RANGE: Newfoundland to Manitoba, south to Oklahoma and Georgia.
ILLINOIS DISTRIBUTION: Rather common in the northern half of Illinois; rare in the southern counties.
The typical and less pubescent variety occurs in the Eastern Hemisphere.
This species is found in more moist situations than any other *Thelypteris* or the related genus *Dryopteris*. It appears to become more robust if not heavily shaded.

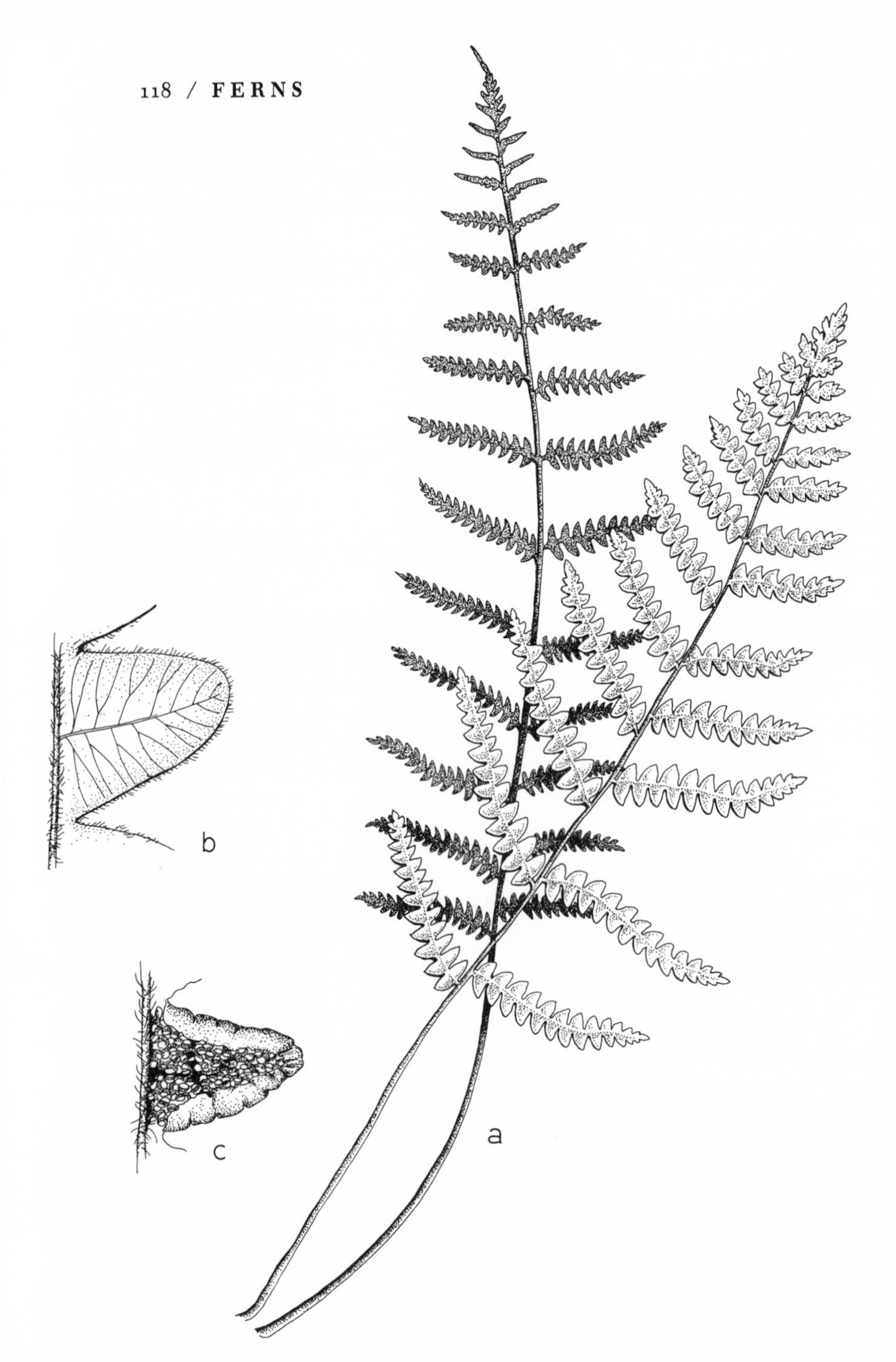

109. *Thelypteris palustris* var. *pubescens* (Marsh Fern). *a.* Leaves, X⅓; *b.* Sterile pinna, X7½; *c.* Fertile pinna, X5.

13. *Dryopteris* ADANS. – Shield Fern

Rhizomes stout; petioles scaly; leaves pinnate-pinnatifid to tripinnate-pinnatifid; sori round; indusium cordate, attached at the center of the sorus.

Those species treated in this volume under the genera *Gymnocarpium* and *Thelypteris* are considered by some botanists as members of *Dryopteris*. *Dryopteris*, as here considered, has evergreen blades, scaly petioles, and a cordate indusium.

The cytology of the species of *Dryopteris* enumerated below has been studied by Walker (1959). His results, along with those of Benedict, Wherry, and Wagner, bear upon the actual relationships of the species.

KEY TO THE SPECIES OF Dryopteris IN ILLINOIS

1. Teeth of leaves bearing short spiny-tips; blades pinnate-pinnatifid to tripinnate.
    2. Indusium eglandular.
        3. Leaves uniform, persisting through the winter __________ 1. *D. carthusiana*
        3. Outer leaves sterile, smaller, spreading, persisting nearly through the winter; inner leaves fertile, larger, erect, not persisting through the winter __________ 5. *D. cristata*
    2. Indusium glandular.
        4. Leaf broadest near middle; spores abortive __________ 2. *D.* × *boottii*
        4. Leaf broadest near base; spores normal or abortive.
            5. Pinnae at right angles to rachis, abruptly narrowed to the long tip; rhizome suberect; spores normal __________ 3. *D. intermedia*
            5. Pinnae ascending from the rachis, gradually narrowed to the tip; rhizome short-creeping; spores abortive __________ 4. *D.* × *triploidea*
1. Teeth of leaves without spiny-tips; blades pinnate-pinnatifid to bipinnate.
    6. Leaves membranous or subcoriaceous; sori not marginal.
        7. Sori borne midway between mid-nerve and margin of leaf segment; lowest pinnules of basal pinnae the longest.
            8. Leaves strongly tapering below, the pinnae at most about 8 cm long __________ 5. *D. cristata*
            8. Leaves moderately tapering below, some or all the pinnae over 8 cm long __________ 6. *D.* × *clintoniana*
        7. Sori borne near midvein of leaf segment; lowest pinnules of basal pinnae shorter than the remainder __________ 7. *D. goldiana*
    6. Leaves coriaceous; sori marginal __________ 8. *D. marginalis*

1. **Dryopteris carthusiana** (Villars) H. P. Fuchs, in Bull. Soc. Bot. France 105:339. 1959. *Fig. 110.*

Polypodium spinulosum O. F. Muell. in Fl. Danica 4 (12):5. 1777, non Burm. f. (1768).

Polypodium carthusianum Villars, in Hist. Pl. Dauph. 1:292. 1786.

Polypodium lanceolatocristatum Hoffm. in Roem. & Ust. Mag. 9:9. 1790.

Aspidium spinulosum (O. F. Muell.) Sw. in Journ. Bot. Schrad. 1800 (2):38. 1801.

Dryopteris spinulosa (O. F. Muell.) Watt, in Can. Nat. II. 3: 159. 1867.

Dryopteris austriaca var. *spinulosa* (O. F. Muell.) Fiori, in Fl. Ital. Crypt. 5:115. 1943.

Dryopteris lanceolatocristata (Hoffm.) Alston, in Watsonia 4 (1):41. 1957.

Usually evergreen from stout, scaly rhizomes; leaves bipinnate to tripinnate, to 75 cm long, the pinnae numerous, firm but not coriaceous, not minutely glandular, the pinnules short spiny-toothed, the lowest pinnule of the basal pinnae the longest; petiole firm, scaly; sori round, small, borne on the back of the leaf segment away from the margin; indusium eglandular, tetraploid; $2n = 164$.

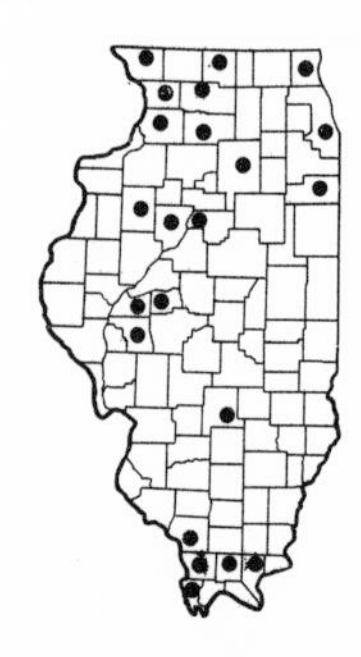

COMMON NAME: Spinulose Woodfern.

HABITAT: Moist, rocky woodlands; rarely in dry situations.

RANGE: Labrador to Alberta, south to Idaho, Missouri, and Virginia.

ILLINOIS DISTRIBUTION: Occasional in the southern, west-central, and northern counties.

The general lack of glands, either on the blades or the indusia, distinguishes this species from the closely related *D.* × *boottii*, *D.* × *triploidea*, and *D. intermedia*. *Dryopteris cristata*, another essentially eglandular species with spiny-tipped teeth on the leaf margins, has dimorphic leaves.

Much disagreement exists concerning the proper disposition of this plant. It and the following three species are sometimes considered to be varieties of the same species. Cytological study indicates the three each should have specific recognition.

The name *Dryopteris spinulosa* is evidently not legitimate, its basionym being a later homonym. It appears that the proper name for this species is *D. carthusiana.*

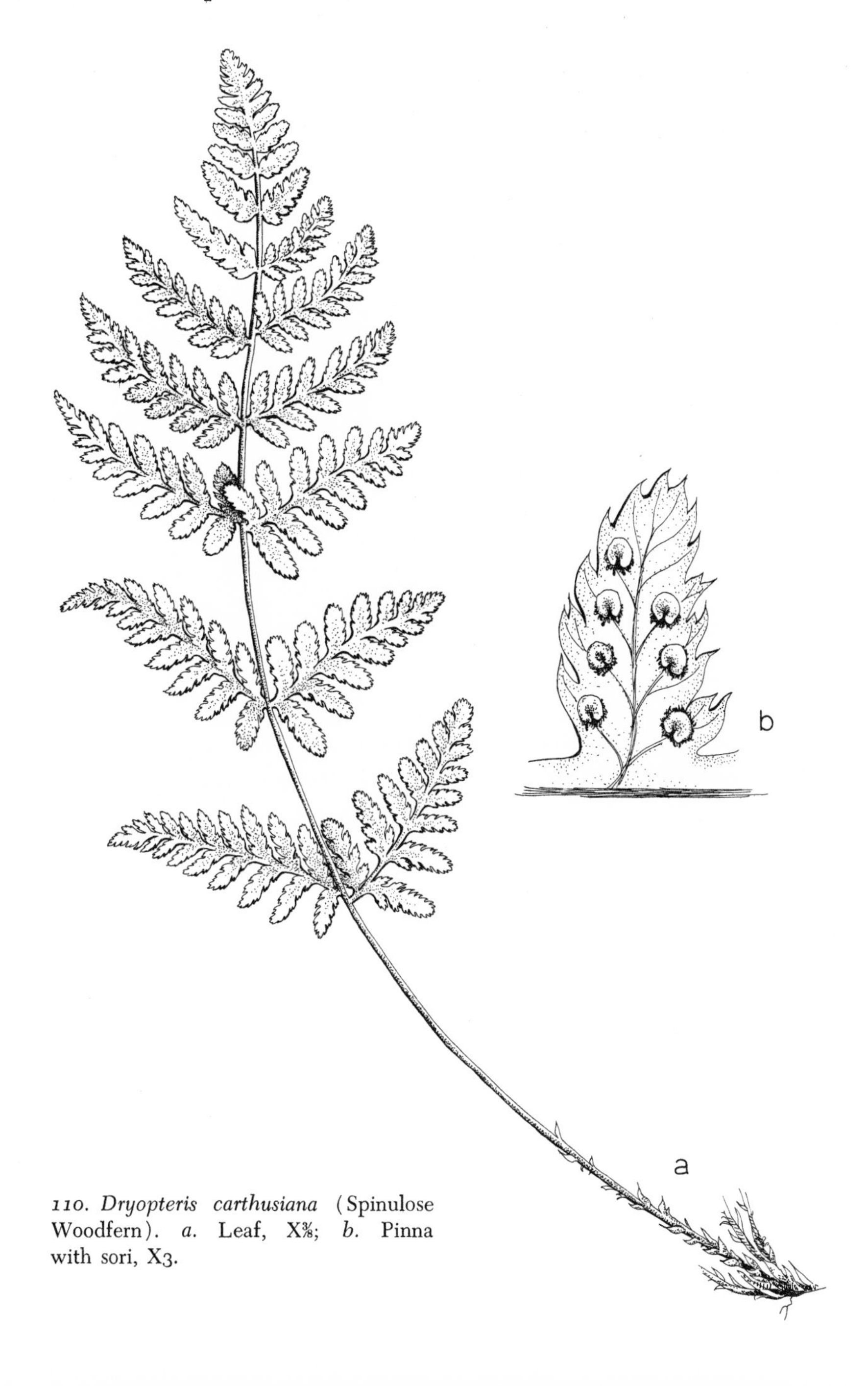

110. Dryopteris carthusiana (Spinulose Woodfern). *a.* Leaf, X⅜; *b.* Pinna with sori, X3.

111. Dryopteris × boottii (Boott's Woodfern). *a.* Leaf, X¼; *b.* Pinna with sori, X2½.

2. Dryopteris × boottii (Tuckerm.) Underw. Our Nat. Ferns 117. 1893. *Fig. 111.*

Aspidium boottii Tuckerm. in Hovey's Mag. Hort. 9:145. 1843.

Aspidium spinulosum var. *boottii* (Tuckerm.) D. C. Eaton in Gray, Man. 665. 1867.

Thelypteris boottii (Tuckerm.) Nieuwland, in Am. Midl. Nat. 1:226. 1910.

Leaves bipinnate-pinnatifid, broadest near the middle, the pinnae firm but not coriaceous, minutely glandular, the pinnules short spiny-toothed; petiole firm, scaly; sori round, borne on the back of the leaf segments away from the margin; indusium minutely glandular; spores abortive; triploid; 2n = 123.

COMMON NAME: Boott's Woodfern.

HABITAT: Moist, rocky woodlands.

RANGE: Nova Scotia to Minnesota, south to Illinois and Virginia.

ILLINOIS DISTRIBUTION: Very rare; known only from La Salle County (deep, mesic ravine, Matthieson State Park, June 13, 1962, *R. H. Mohlenbrock 14973*).

This plant, rather difficultly distinguishable from *D. intermedia*, is a sterile hybrid between the diploid *D. intermedia* and the tetraploid *D. cristata*, as first proposed by Dowell (1908).

3. Dryopteris intermedia (Muhl.) Gray, Man. 630. 1848. *Fig. 112.*

Aspidium intermedium Muhl. ex Willd. Sp. Pl. 5:262. 1810.

Aspidium spinulosum var. *intermedium* (Muhl.) D. C. Eaton in Gray, Man. 665. 1867.

Dryopteris spinulosa var. *intermedia* (Muhl.) Underw. Our Nat. Ferns 116. 1893.

Nephrodium spinulosum var. *intermedium* (Muhl.) Davenp. ex Gilb. List N. Am. Pterid. 17. 1901.

Filix-mas spinulosa var. *intermedia* (Muhl.) Farw. in Am. Midl. Nat. 12:257. 1931.

Dryopteris austriaca var. *intermedia* (Muhl.) Morton, in Am. Fern Journ. 40:222. 1950.

Evergreen from stout, suberect rhizomes; leaves bipinnate to tripinnate, to 75 cm long, the pinnae numerous, at right angles

112. Dryopteris intermedia (Common Woodfern). *a.* Leaf, X½ (shortened); *b.* Pinna with sori, X5; *c.* Sorus, X30.

to the rachis, firm but not coriaceous, minutely glandular, the pinnules short spiny-toothed, the lowest pinnule of the basal pinnae equalling or shorter than the pinnule immediately above; petioles and rachis glandular, scaly; sori round, small, borne on the back of the leaf segment away from the margin; indusium minutely glandular; diploid; 2n = 82.

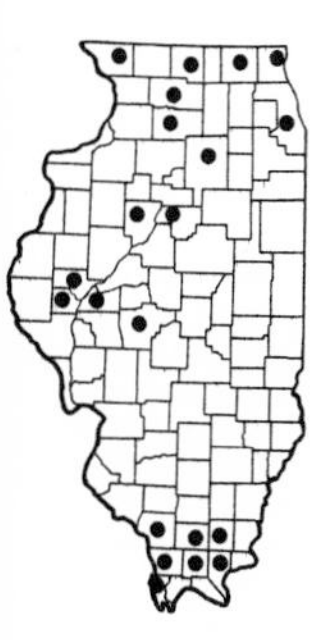

COMMON NAME: Common Woodfern; Intermediate Fern.
HABITAT: Moist, rocky ravines and woodlands.
RANGE: Newfoundland to Ontario, south to Iowa and Georgia; reported by Broun (1938) from Montana.
ILLINOIS DISTRIBUTION: Occasional or rare in the northern half of Illinois and the southern tip of Illinois.

This is the laciest fern in Illinois. The degree of leaf-cutting is rather variable. Although this species is distinguished from *D. carthusiana* mainly by having minutely glandular leaves and indusia, these glands frequently are difficult to find late in the season.

This taxon is considered by some to be a variety of *D. carthusiana*, but cytological evidence indicates otherwise.

4. **Dryopteris × triploidea** Wherry, in Am. Fern Journ. 50:90. 1960. *Fig. 113.*

Nephrodium spinulosum var. *fructuosum* Gilb. List N. Am. Pterid. 37. 1901.

Thelypteris spinulosa var. *fructuosa* (Gilb.) Fern. in Rhodora 28:146. 1926.

Dryopteris spinulosa var. *fructuosa* (Gilb.) Trudell, in Am. Fern Journ. 19:136. 1929.

Dryopteris intermedia var. *fructuosa* (Gilb.) Wherry, Guide to Eastern Ferns 125. 1937.

Dryopteris intermedia f. *fructuosa* (Gilb.) Clute, Our Ferns 183. 1938.

Dryopteris austriaca var. *fructuosa* (Gilb.) Morton, in Am. Fern Journ. 40:222. 1950.

Evergreen from short-creeping rhizomes; leaves bipinnate to tripinnate, to 80 cm long, the pinnae numerous, firm but not coriaceous, minutely glandular, the pinnules short spiny-toothed, the lowest pinnule of the basal pinnae usually equalling or shorter than the pinnule immediately above; petiole and rachis glandular, scaly; sori round, small, borne on the back of the leaf segments away from the margin; indusium minutely glandular; spores abortive; triploid; 2n = 123.

COMMON NAME: Woodfern.

HABITAT: Moist, rocky woodlands.

RANGE: Newfoundland to Ontario, south to Illinois, Tennessee, and Virginia.

ILLINOIS DISTRIBUTION: Very rare; known only from Lake County (damp woods near Barrington, June 14, 1962, *R. H. Mohlenbrock 14988*).

The single locality for this fern in Illinois makes it one of the rarest ferns in the state.

Walker (1959) has demonstrated this fern to be a sterile triploid, the result of crossing the diploid *D. intermedia* and the tetraploid *D. carthusiana.*

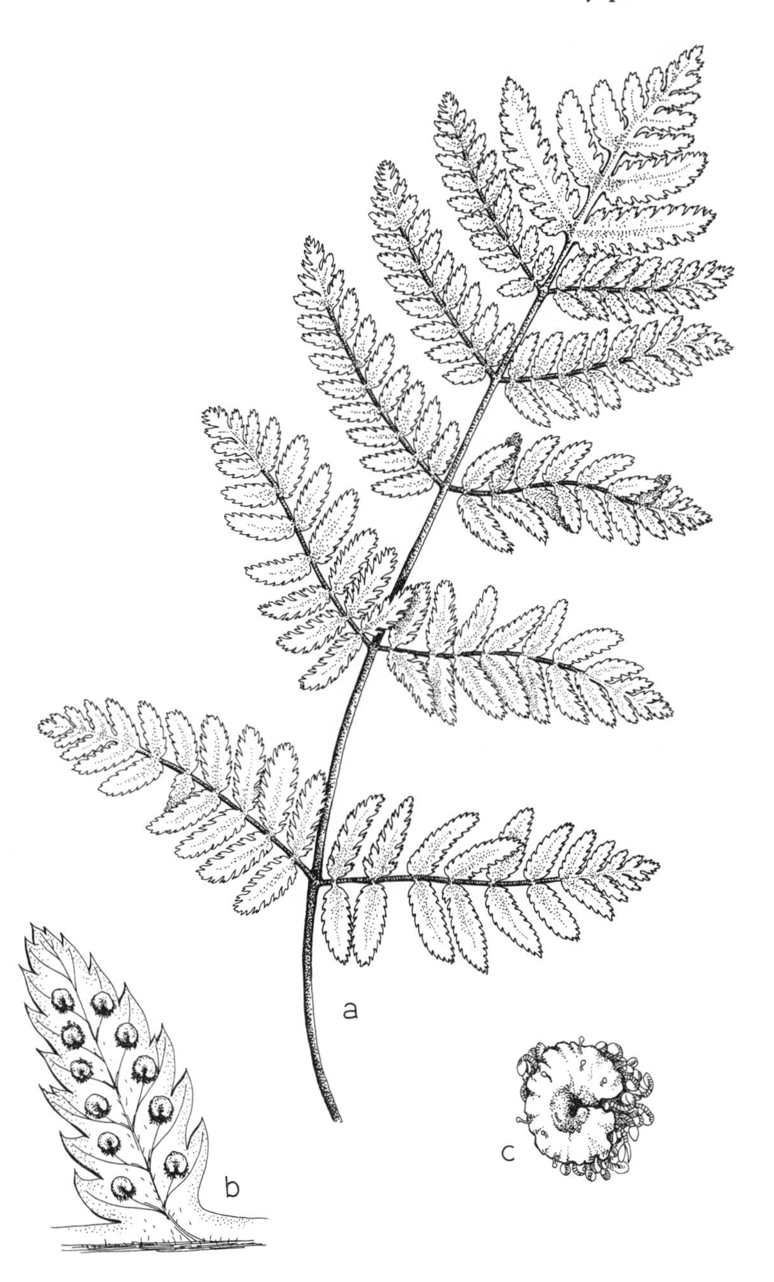

113. Dryopteris × triploidea (Woodfern). *a.* Leaf, X⅜; *b.* Pinna with sori, X3; *c.* Sorus, X30.

114. Dryopteris cristata (Crested Fern). *a.* Leaves, X½; *b.* Pinna with sori, X3; *c.* Sorus, X30.

5. **Dryopteris cristata** (L.) Gray, Man. 631. 1848. *Fig. 114.*

Polypodium cristatum L. Sp. Pl. 1090. 1753.

Aspidium cristatum (L.) Sw. in Journ. Bot. Schrad. 1800 (2):37. 1801.

Nephrodium cristatum (L.) Michx. Fl. Bor. Am. 2:269. 1803.

Thelypteris cristata (L.) Nieuwland, in Am. Midl. Nat. 1:226. 1910.

Rhizomes short-creeping; leaves pinnate-pinnatifid to bipinnate, to 75 cm long, dimorphic, the outer sterile leaves smaller, spreading, tending to persist longer during the winter, the pinnules usually spinulose-toothed; inner fertile leaves taller, erect, not persistent during the winter, the pinnae not more than 8 cm long; petiole stout, scaly; sori round, borne on the back of the leaf segments away from the margin; indusium eglandular; tetraploid; 2n = 246.

COMMON NAME: Crested Fern.

HABITAT: Low, moist woodlands.

RANGE: Newfoundland to Alberta, south to Idaho, Nebraska, Arkansas, and Virginia; Europe.

ILLINOIS DISTRIBUTION: Occasional in the eastern counties of the northern one-third of the state; absent elsewhere. This beautiful fern, when seen in the field, exhibits its dimorphism well. The outer "whorl" of sterile leaves is smaller and more spreading than the taller, more erect inner cluster of leaves.

6. **Dryopteris × clintoniana** (D. C. Eaton) Dowell, in Proc. Staten Is. Assoc. 1:64. 1906. *Fig. 115.*

Aspidium cristatum var. *clintonianum* D. C. Eaton in Gray, Man. 665. 1867.

Dryopteris cristata var. *clintoniana* (D. C. Eaton) Underw. Our Nat. Ferns 115. 1893.

Thelypteris cristata var. *clintoniana* (D. C. Eaton) Weatherby, in Rhodora 21:177. 1919.

Rhizome short-creeping; leaves pinnate-pinnatifid, to 90 cm long, not dimorphic, with 10–15 pairs of pinnae, the pinnae serrate but not spinulose, eglandular, some or all over 8 cm long; petiole stout, scaly; sori round, borne on the back of the leaf segments away from the margin; indusium eglandular; hexaploid; 2n = 246.

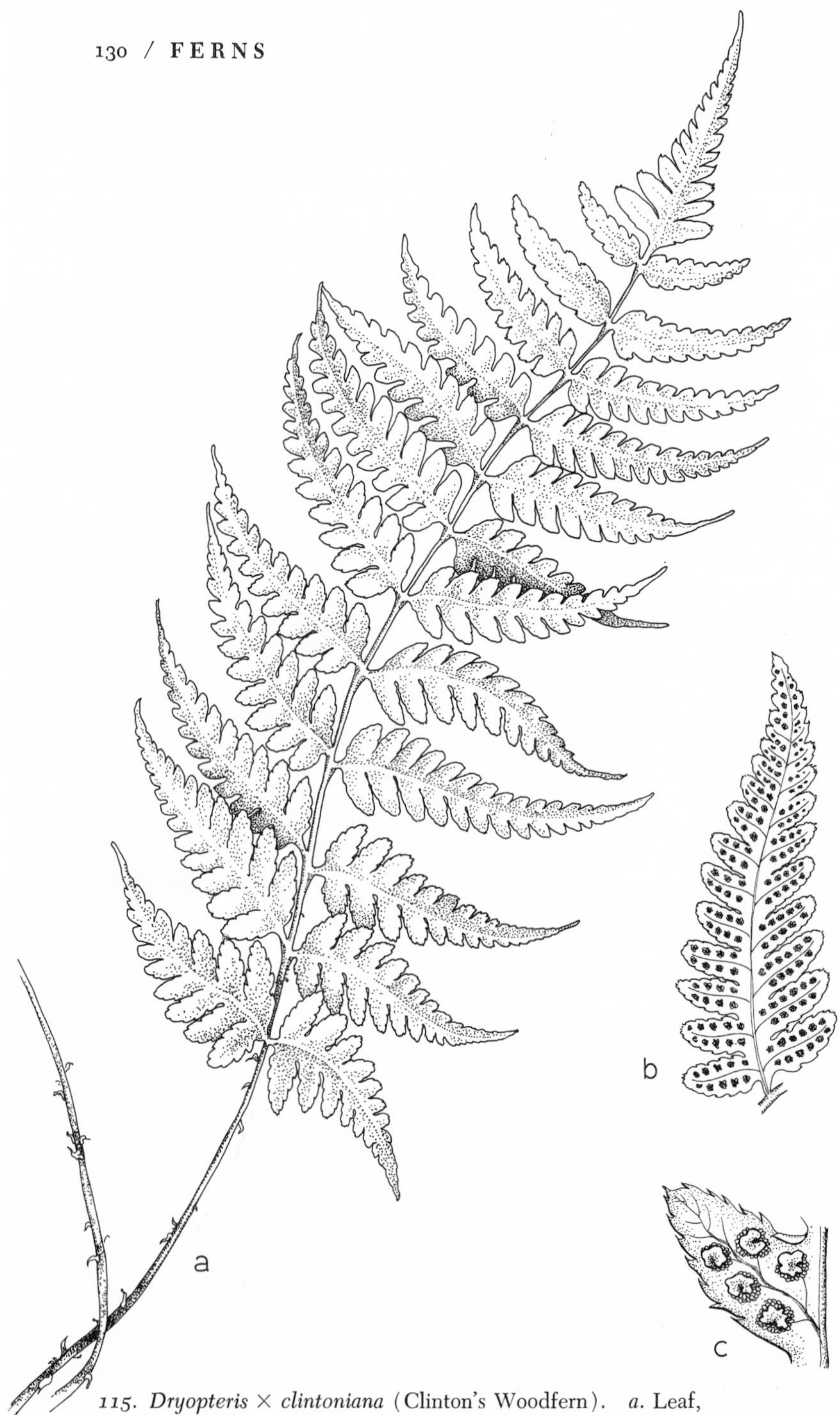

115. *Dryopteris × clintoniana* (Clinton's Woodfern). *a.* Leaf, X¼; *b.* Fertile pinna, X½; *c.* Pinna with sori, X1¼.

COMMON NAME: Clinton's Woodfern.
HABITAT: Moist, rich woodlands.
RANGE: Quebec to Ontario, south to Missouri and Georgia.
ILLINOIS DISTRIBUTION: Very rare; known only from St. Clair County (near Belleville, August 7, 1870, collector unknown) before the turn of the century; probably now extinct in Illinois.

Some authors consider this fern to be a variety of *D. cristata*, while others believe it to be an allopolyploid (hexaploid) of *D. cristata* and *D. goldiana*. Cytological evidence presented by Walker (1959) bears out this latter premise. The larger size and the lack of dimorphism distinguish this species from *D. cristata*.

7. **Dryopteris goldiana** (Hook.) Gray, Man. 631. 1848. *Fig. 116.*

Aspidium goldianum Hook. ex Goldie, in Edinb. Philos. Journ. 6:333. 1822.
Nephrodium goldianum (Hook.) Hook. & Grev. Icon. Fil. Pl. 102. 1829.
Thelypteris goldiana (Hook.) Nieuwland, in Am. Midl. Nat. 1:226. 1910.

Rhizome stout, scaly, short-creeping; leaves dark green, pinnate-pinnatifid to bipinnate, to nearly 1 m long, with up to 35 pairs of pinnae, the pinnae eglandular, with the ultimate segments shallowly toothed; petiole pale brown, with large, blackish scales at base; sori round, borne on the back of the leaf segment near the midvein; indusium eglandular; diploid; 2n = 82.

COMMON NAME: Goldie's Fern.
HABITAT: Moist, shaded woodlands.
RANGE: New Brunswick to Ontario, south to Iowa, Tennessee, and North Carolina.
ILLINOIS DISTRIBUTION: Rare; known from seven northern counties and Jackson County.

116. Dryopteris goldiana (Goldie's Fern). *a.* Leaf, X¼; *b.* Pinna with sori, X1¼.

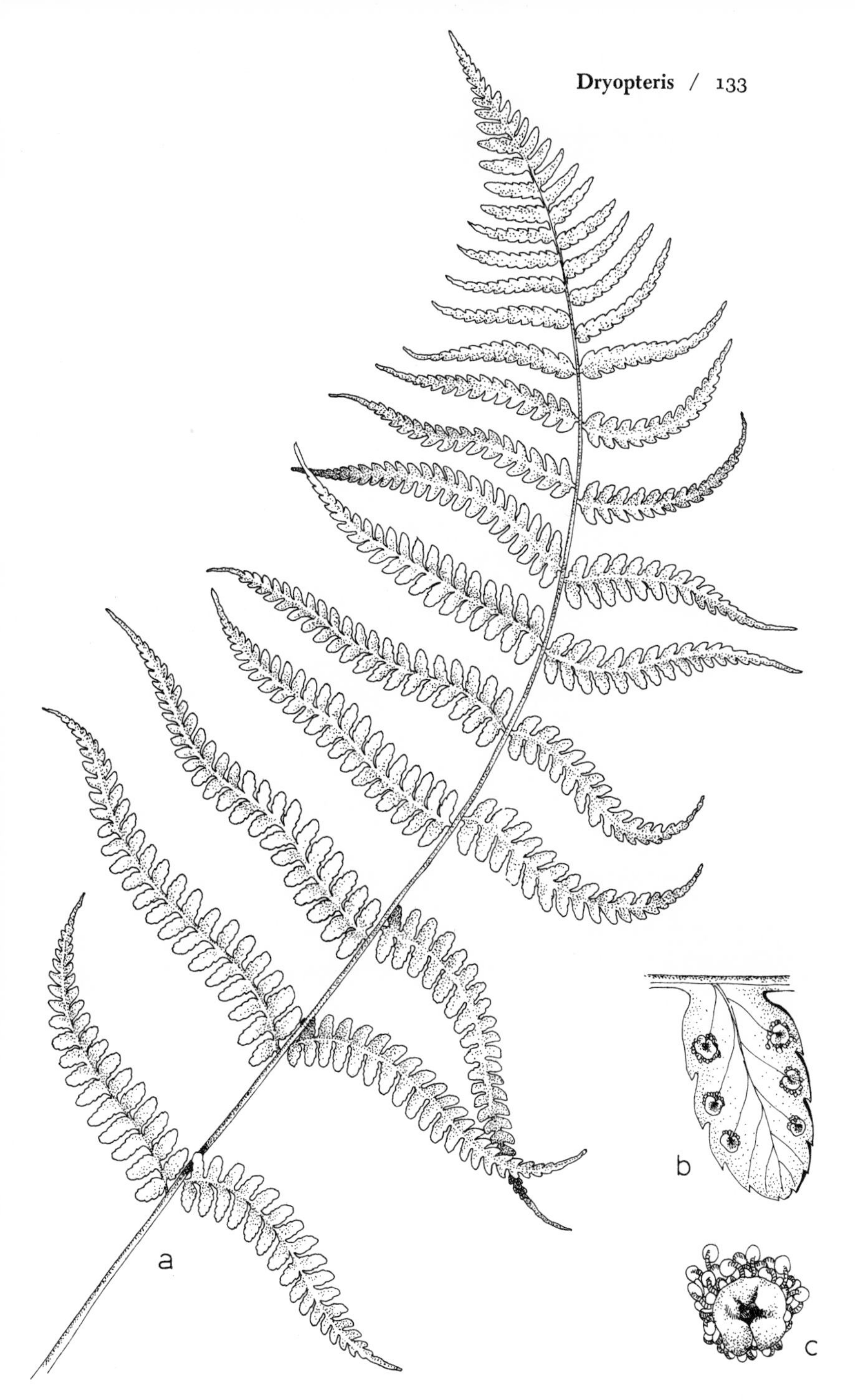

117. Dryopteris marginalis (Marginal Fern). *a.* Leaf, X½; *b.* Pinna with sori, X4; *c.* Sorus, X20.

8. Dryopteris marginalis (L.) Gray, Man. 632. 1848. *Fig. 117.*

Polypodium marginale L. Sp. Pl. 1091. 1753.

Aspidium marginale (L.) Sw. in Journ. Bot. Schrad. 1800 (2):35. 1801.

Nephrodium marginale (L.) Michx. Fl. Bor. Am. 2:267. 1803.

Thelypteris marginalis (L.) Nieuwland, in Am. Midl. Nat. 1:226. 1910.

Evergreen from stout, scaly rhizómes; leaves bipinnate to bipinnate-pinnatifid, to 60 cm long, with up to 20 pairs of pinnae, the pinnae coriaceous, glabrous, eglandular, entire or shallowly toothed; petiole stout, scaly (at least below); sori round, borne on the back of the leaf segment at the margin; indusium centrally attached, eglandular; diploid; 2n = 82.

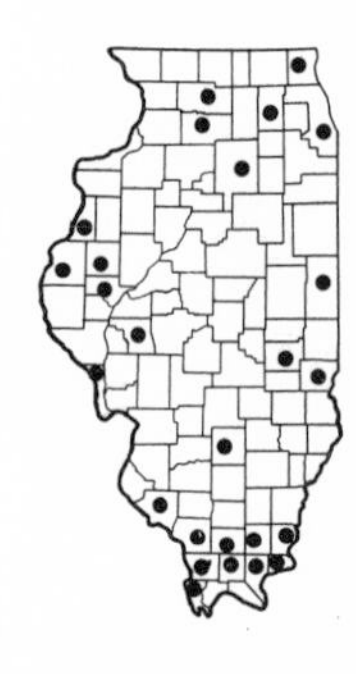

COMMON NAME: Marginal Fern; Leather Fern; Marginal Shield Fern.

HABITAT: Rocky woodlands.

RANGE: Nova Scotia to British Columbia, south to Oklahoma and Georgia.

ILLINOIS DISTRIBUTION: Common in the southern tip of Illinois; occasional to rare elsewhere.

This handsome, evergreen fern has deep green leaves which may even appear to be blue-green. Some variation occurs as to amount of leaf-cutting of the ultimate segments. The crested f. *davenportii* has been found in Pope County (Dixon Springs State Park). The marginal sori distinguish it from all other species of *Dryopteris.*

14. Woodwardia SM. – Chain Fern

Rhizomes long, creeping, scaly; sterile and fertile leaves (in our species) similar, pinnate-pinnatifid; sori slender, borne parallel to the main veins, covered by an indusium.

Only the following species occurs in Illinois.

1. Woodwardia virginica (L.) Sm. in Mem. Acad. Turin 5:412. 1793. *Fig. 118.*

Blechnum virginicum L. Mant. 2:307. 1771.

Anchistea virginica (L.) Presl, Epim. Bot. 71. 1851.

Deciduous perennial with rather stout, creeping, scaly rhizomes; leaves pinnate-pinnatifid, to 12 (–15) dm long, with up to 20 pairs of pinnae, the pinnae ovate to narrowly ovate, sessile,

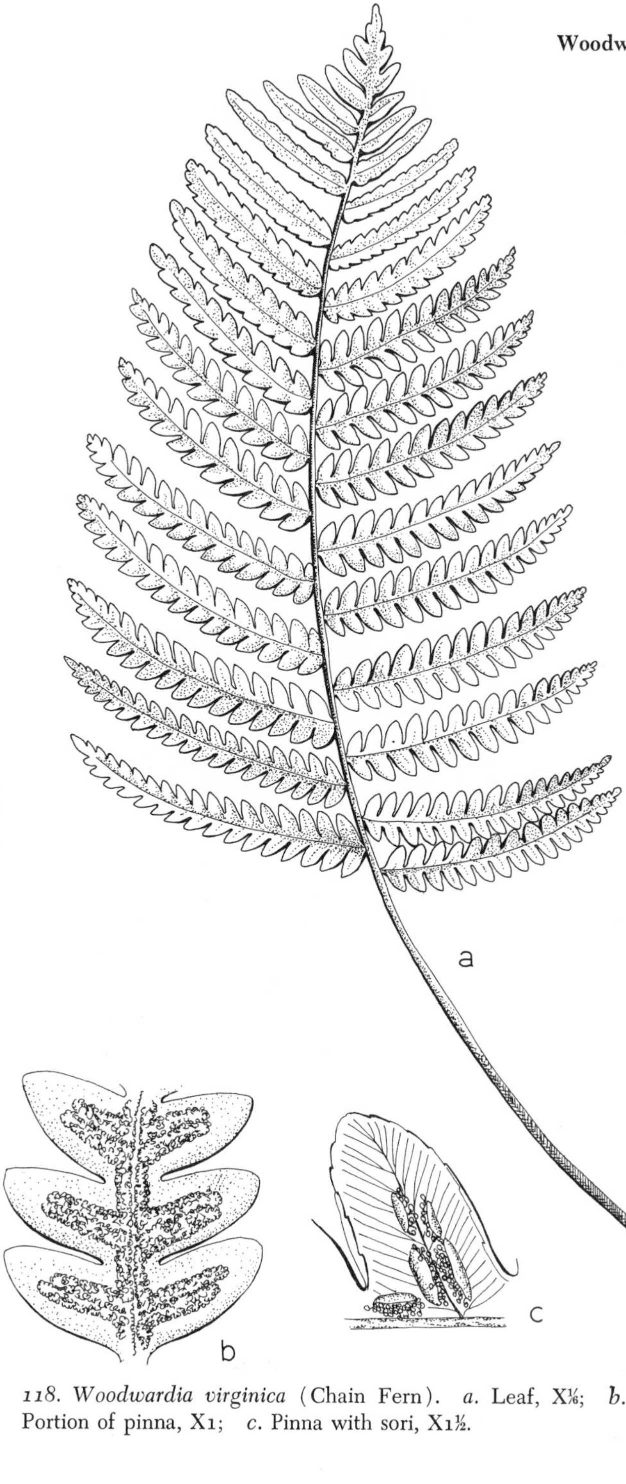

118. *Woodwardia virginica* (Chain Fern). *a.* Leaf, X⅙; *b.* Portion of pinna, X1; *c.* Pinna with sori, X1½.

glabrous; petiole purplish-brown, shiny, glabrous; sori elongate, confluent at maturity, parallel in a single row to the midnerve; indusium laterally attached.

COMMON NAME: Chain Fern.

HABITAT: Bogs.

RANGE: Nova Scotia to Ontario, south to Texas and Florida; Bermuda.

ILLINOIS DISTRIBUTION: Very rare; known only from Lake County (tamarack bog near Antioch, *G. D. Fuller & G. N. Jones 16512,* in 1944) in extreme northeastern Illinois.

This fern takes its common name from the chain-like rows of elongate sori. It was unknown from Illinois until 1944.

A peculiar feature is the single row of areolae formed by the veins on either side of the main veins.

15. *Athyrium* ROTH – Lady Ferns

Rhizomes rather stout, creeping; leaves deciduous, pinnate to tripinnate; sori elongate, straight or curved; indusium attached laterally.

KEY TO THE SPECIES OF Athyrium IN ILLINOIS

1. Blades once-pinnate; pinnae entire............1. *A. pycnocarpon*
1. Blades pinnate-pinnatifid to bi- (tri-)pinnate; pinnae pinnatifid.
    2. Blades pinnate-pinnatifid; indusium light brown (at maturity), firm, parallel to the veins; sori straight or nearly so.......................................2. *A. thelypterioides*
    2. Blades at least bipinnate; indusium dark brown (at maturity), membranous, many crossing the veins; sori more or less curved ....3. *A. filix-femina*

1. Athyrium pycnocarpon (Spreng.) Tidestrom, Elys. Marianum 1:36. 1906. *Fig. 119.*

Asplenium angustifolium Michx. Fl. Bor. Am. 2:265. 1803, non Jacq. (1786).

Asplenium pycnocarpon Spreng. in Anleit. Kennt. Gewächse 3:112. 1804.

Diplazium angustifolium (Michx.) Butters, in Rhodora 19:178. 1917.

Diplazium pycnocarpon (Spreng.) Broun, Index N. Am. Ferns 60. 1938.

Tall, delicate, deciduous perennial from stout rhizomes; leaves once-pinnate, glabrous, to nearly 1 m long, with up to 40 pairs of pinnae, the sterile broader and thinner, the fertile narrower and firmer; petioles green to pale brown, glabrous; sori elongated, straight, borne on the back of the pinnae; indusium attached laterally.

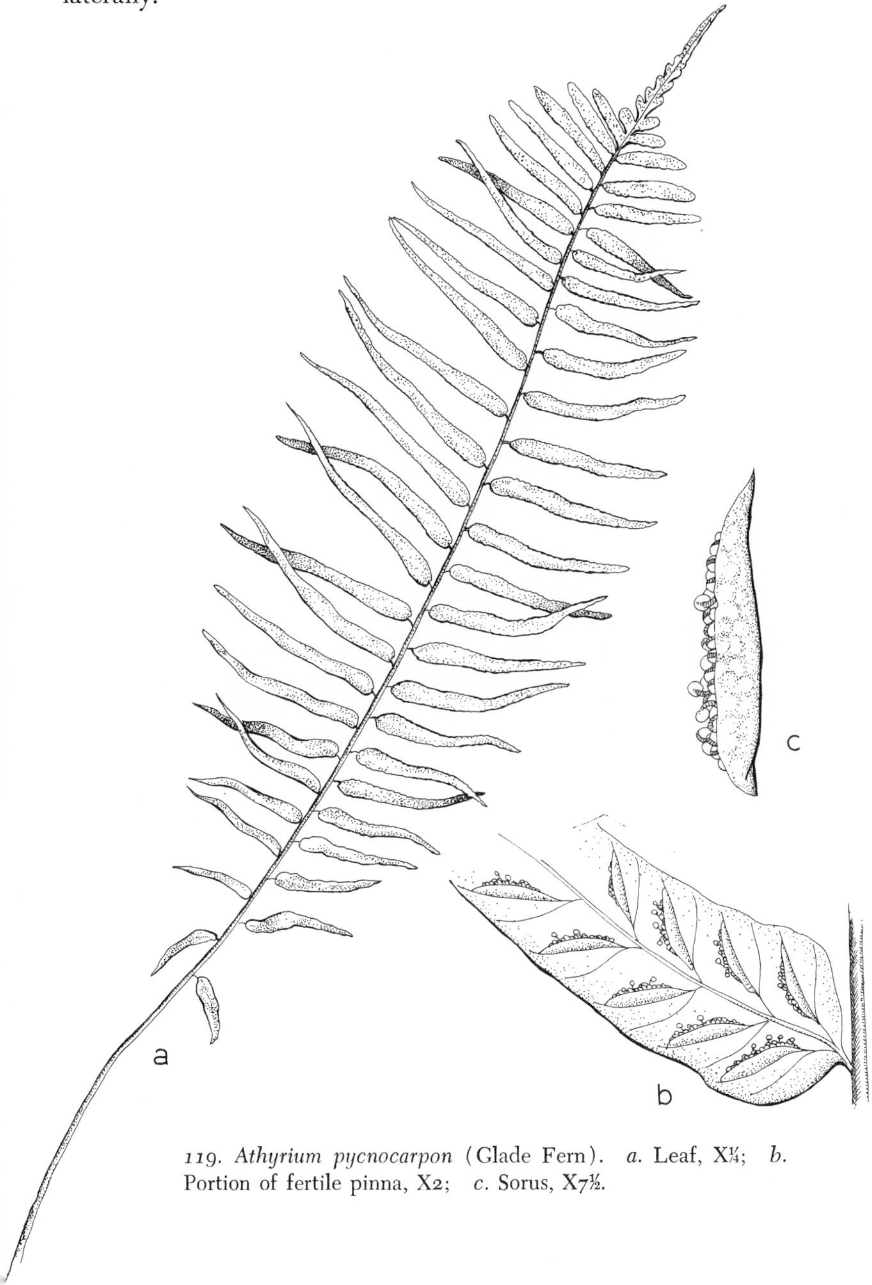

119. *Athyrium pycnocarpon* (Glade Fern). *a.* Leaf, X¼; *b.* Portion of fertile pinna, X2; *c.* Sorus, X7½.

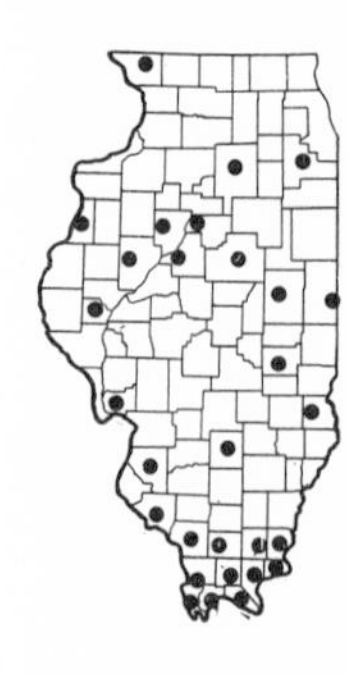

COMMON NAME: Narrow-leaved Spleenwort; Glade Fern.
HABITAT: Moist, shaded woodlands.
RANGE: Quebec to Minnesota, south to Kansas, Louisiana, and Georgia.
ILLINOIS DISTRIBUTION: Occasional in the southern three-fourths of Illinois; also in JoDaviess County.
The elongated, delicate, once-pinnate leaves distinguish this handsome fern.

2. **Athyrium thelypterioides** (Michx.) Desv. in Mem. Soc. Linn. Paris 6:266. 1827. *Fig. 120.*

Asplenium acrostichoides Sw. in Journ. Bot. Schrad. 1800 (2):54. 1801.
Asplenium thelypterioides Michx. Fl. Bor. Am. 2:265. 1803.
Diplazium thelypterioides (Michx.) Presl, Tent. Pterid. 114. 1836.
Athyrium acrostichoides (Sw.) Diels in Engler & Prantl, Nat. Pflanzenf. 1 (2):223. 1899, non A. *acrostichoideum* Bory ex Merat (1836).
Diplazium acrostichoides (Sw.) Butters, in Rhodora 19:178. 1917.

Rather large, deciduous perennial from stout rhizomes; leaves pinnate-pinnatifid, to nearly 1 m long, with up to 30 pairs of pinnae, the pinnae long-tapering, pinnatifid, glabrous except along the mid-nerve below, the ultimate segments entire or denticulate; petiole and rachis chaffy; sori elongated, straight or slightly curved, borne in two rows on the back of the leaf segments; indusium silvery at first, attached laterally.

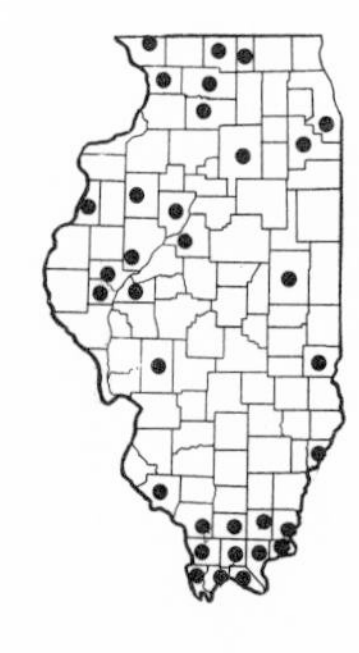

COMMON NAME: Silvery Spleenwort.
HABITAT: Rich woodlands.
RANGE: Quebec to Ontario, south to Louisiana and Georgia; Asia.
ILLINOIS DISTRIBUTION: Occasional throughout Illinois; apparently absent from the south-central counties.
This attractive fern derives its common name from the silvery appearance of the young indusia.

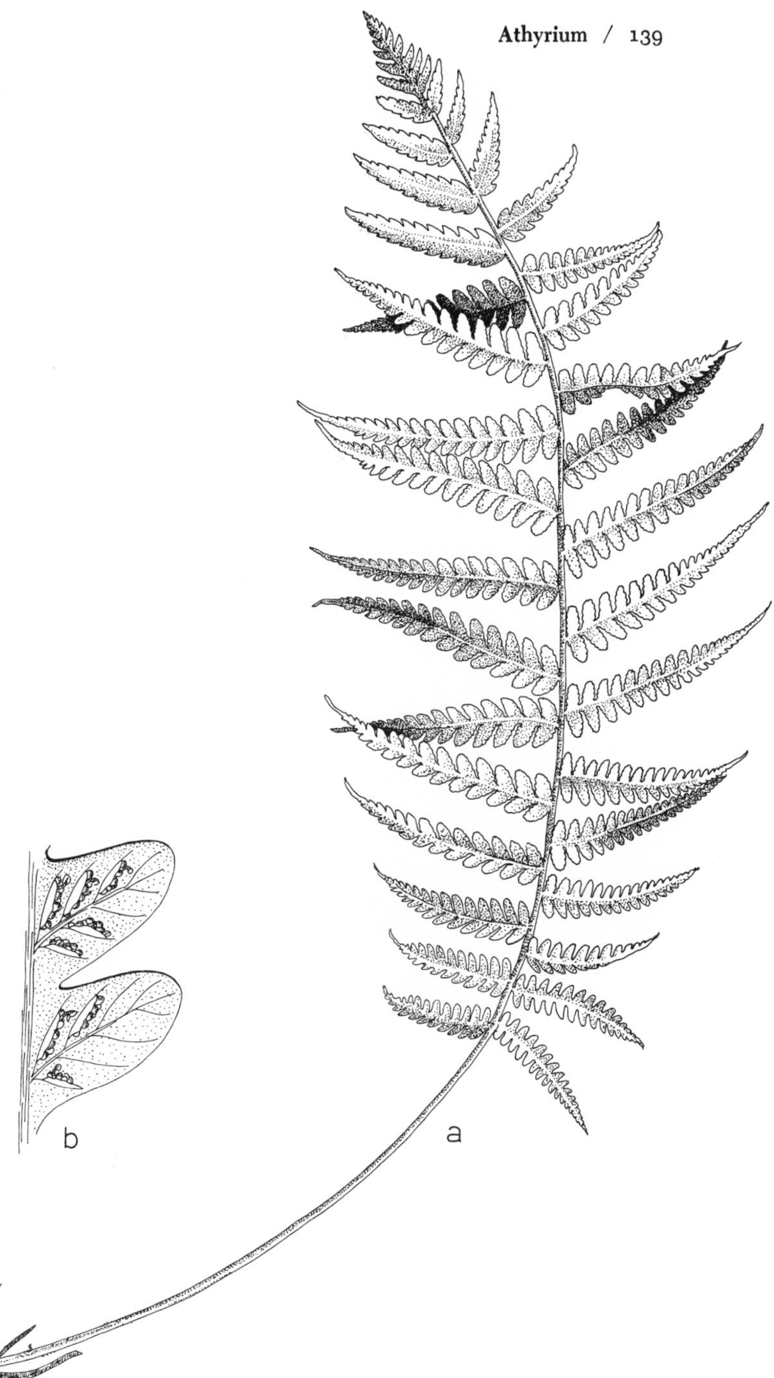

120. Athyrium thelypterioides (Silvery Spleenwort). *a.* Leaf, X⅓; *b.* Pinnae with sori, X3½.

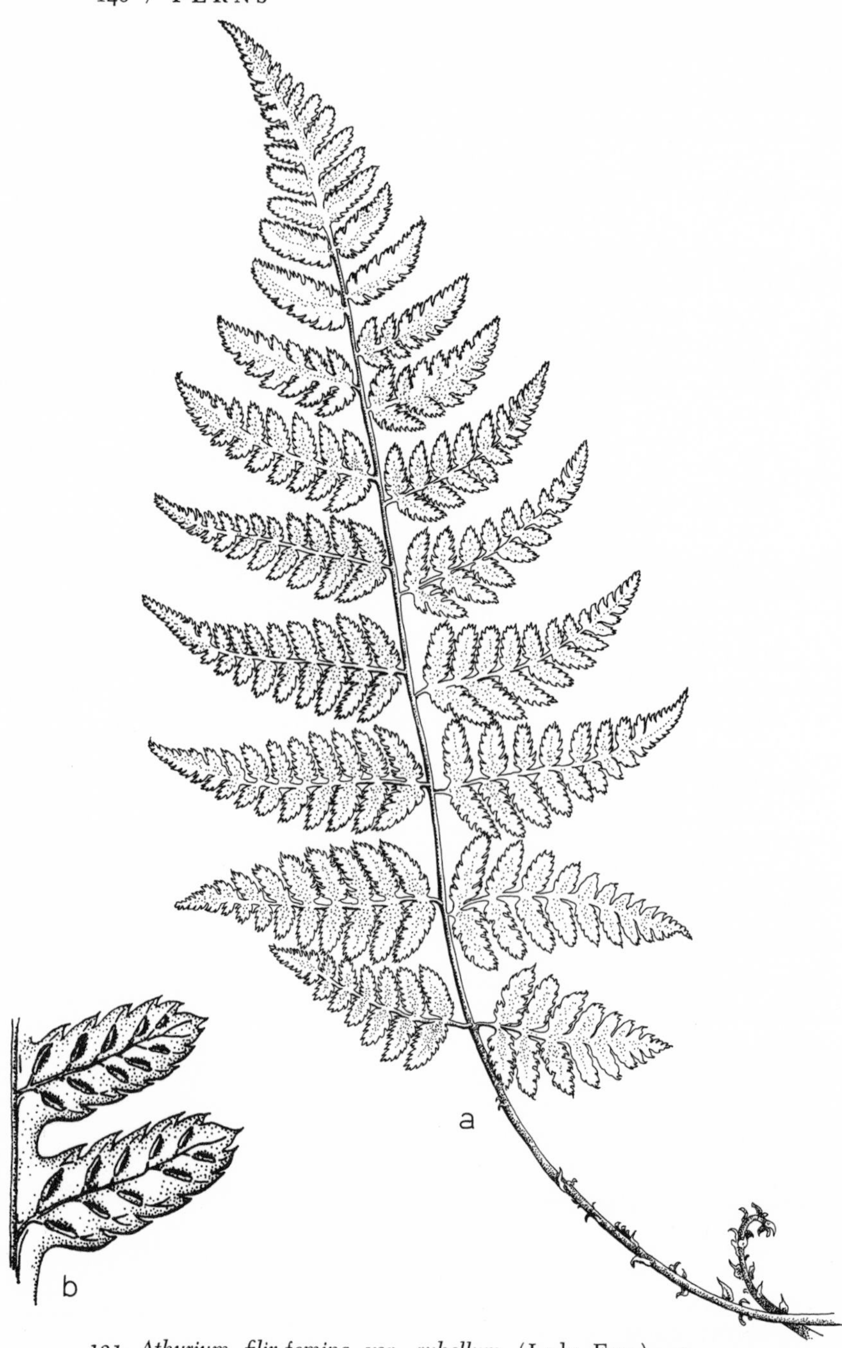

121. Athyrium filix-femina var. *rubellum* (Lady Fern). *a.* Leaf, X¼; *b*. Fertile pinnae, X1½.

3. **Athyrium filix-femina** (L.) Roth in Roem. in Arch. Bot. 2 (1):106. 1799.

Two varieties of this species occur in Illinois, distinguishable by the following key.

a. Fourth or fifth pair of pinnae (from the base) the largest; rachis glandular; indusium eglandular; petiole about one-half as long as the blade____________________3a. *A. filix-femina* var. *rubellum*

a. Second or third pair of pinnae (from the base) the largest; rachis eglandular; indusium glandular; petiole about as long as the blade __________________________3b. *A. filix-femina* var. *asplenioides*

3a. **Athyrium filix-femina** (L.) Roth var. **rubellum** Gilb. List N. Am. Pterid. 35. 1901. *Fig. 121.*

Aspidium angustum Willd. Sp. Pl. 5:277. 1810.
Athyrium angustum (Willd.) Presl, Rel. Haenk. 1:39. 1825.
Asplenium michauxii Spreng. Syst. 4:88. 1827.
Athyrium elatius Link, Fil. Sp. Hort. Berol. 94. 1841.
Athyrium filix-femina var. *michauxii* (Spreng.) Farw. in Ann. Rep. Mich. Acad. Sci. 18:79. 1916.
Athyrium angustum var. *elatius* (Link) Butters, in Rhodora 19:191. 1917.
Athyrium angustum var. *rubellum* (Gilb.) Butters, in Rhodora 19:193. 1917.
Athyrium filix-femina f. *rubellum* (Gilb.) Farw. in Papers Mich. Acad. Sci. 2:13. 1923.
Athyrium filix-femina f. *elatius* (Link) Clute, Our Ferns 224. 1938.

Tall, handsome, deciduous perennial from short, stout, scaly rhizomes; leaves bipinnate, to nearly 1.5 m long, the pinnae broadest near base, membranous, the pinnules shallowly lobed or serrate; petioles slender, about one-half as long as the blade, chaffy near base or glabrous, not scaly; rachis glandular; sori elongated, curved, distinct or confluent, borne on the back of the leaf segments; indusium ciliate but not glandular, attached laterally; spores yellow-brown.

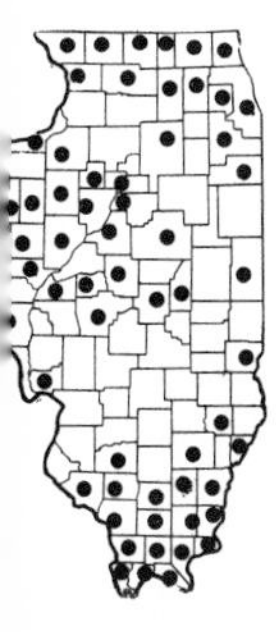

COMMON NAME: Lady Fern.

HABITAT: Moist, open woodlands and borders of swamps.

RANGE: Newfoundland to Manitoba, south to Colorado, Missouri, and Virginia.

ILLINOIS DISTRIBUTION: Rather common throughout the state.

This beautiful fern is highly variable in manner of leaf-cutting. Many of the variations given epithets by Butters (1917) and Weatherby (1936) do not occur in Illinois and seem scarcely justifiable.

The sori may be confluent at maturity, or they may remain distinct. Specimens of the first type may be known as f. *elatius*, while those of the latter have been called f. *rubellum*. Both occur together throughout Illinois. They are not distinguished on the distribution map.

Lady fern might be confused with *Dryopteris carthusiana* or *D. intermedia*, but these last species are spinulose-toothed and bear numerous scales on the petiole.

3b. Athyrium filix-femina (L.) Roth var. **asplenioides** (Michx.) Farw. in Papers Mich. Acad. Sci. 2:13. 1923. *Fig. 122.*

Nephrodium asplenioides Michx. Fl. Bor. Am. 2:268. 1803.
Athyrium asplenioides (Michx.) Desv. Prod. 266. 1827.
Asplenium asplenioides (Michx.) D. C. Eaton ex Chapm. Fl. So. U. S. 593. 1860.

Tall, handsome, deciduous perennial from elongated rhizomes with few or no persistent bases of old stipes; fronds bipinnate, to nearly 1.5 m long, the second or third pair of pinnae from the base the largest, membranous, the pinnules shallowly lobed; petioles slender, about as long as the blade, chaffy near base or glabrous, not scaly; rachis eglandular; sori elongated, curved, distinct or confluent, borne on the back of the leaf segments; indusium glandular-ciliate, attached laterally; spores blackish.

COMMON NAME: Southern Lady Fern.

HABITAT: Moist woodlands.

RANGE: Massachusetts to Oklahoma, south to Texas and Florida.

ILLINOIS DISTRIBUTION: Very rare; known only from Jackson County (Giant City State Park, damp, shady sandstone ledge in woods, September 18, 1941, *R. Tryon 4649*).

Although some workers prefer to treat this as a species, there are so many intermediate specimens between this and var. *rubellum* that specific segregation does not seem justifiable.

122. *Athyrium filix-femina* var. *asplenioides* (Southern Lady Fern). *a*. Leaf, X½; *b*. Pinna with sori, X6.

16. *Asplenium* L. – Spleenwort

Rootstocks small; leaves usually evergreen, simple and entire to bipinnate; sori elongated; indusium attached laterally.

Asplenium is an interesting genus of rather small ferns in which much hybridization occurs. For an account of the hybrid nature of various taxa, see Wagner (1954).

KEY TO THE SPECIES OF Asplenium IN ILLINOIS

1. Leaves simple, unlobed (rarely a single pair of lobe-like auricles at base); veins forming a network____________1. *A. rhizophyllum*
1. Leaves simple and pinnatifid, or pinnate to bipinnate-pinnatifid; veins free.
    2. Rachis green throughout.
        3. Petiole green throughout_______________2. *A. ruta-muraria*
        3. Petiole brown at base, sometimes throughout.
            4. Blades entirely pinnatifid, or with merely the lowest pair of pinnae distinct; spores normal______3. *A. pinnatifidum*
            4. Blades pinnatifid above, pinnate below for at least two pairs of pinnae; spores abortive.
                5. Petiole brown throughout__________4. *A.* × *gravesii*
                5. Petiole brown at base, green above____5. *A.* × *trudellii*
    2. Rachis partly or entirely brown, or black.
        6. Rachis brown below (usually about half its length), green above.
            7. Blades once-pinnate below, pinnatifid above; spores abortive___________________________6. *A.* × *kentuckiense*
            7. Blades bipinnate to bipinnate-pinnatifid; spores normal____7. *A. bradleyi*
        6. Rachis brown or black throughout, or at least for three-fourths the length.
            8. Rachis brown for three-fourths its length, green above; blade pinnatifid for half its length, the tip narrowly caudate_________________________8. *A.* × *ebenoides*
            8. Rachis brown or black throughout; blade pinnatifid only at apex, the tip not caudate.
                9. Pinnae not auriculate____________9. *A. trichomanes*
                9. Pinnae auriculate at base of upper margin.
                    10. Pinnae pairs opposite (sometimes alternate); fertile and sterile leaves similar; auricle of pinna not overlapping rachis______________10. *A. resiliens*
                    10. Pinnae pairs alternate; fertile leaves more erect and taller than sterile leaves; auricle of pinna overlapping rachis______________11. *A. platyneuron*

1. Asplenium rhizophyllum L. Sp. Pl. 1078. 1753. *Fig. 123.*

Camptosorus rhizophyllus (L.) Link, Hort. Berol. 2:69. 1833. Evergreen from slender, scaly rhizomes; leaves simple, entire or undulate, occasionally broadly auriculate at base, subcoriaceous, glabrous, lance-linear, tapering to an exaggerated, rooting tip, the veins nearer the midnerve anastomosing, those nearer the

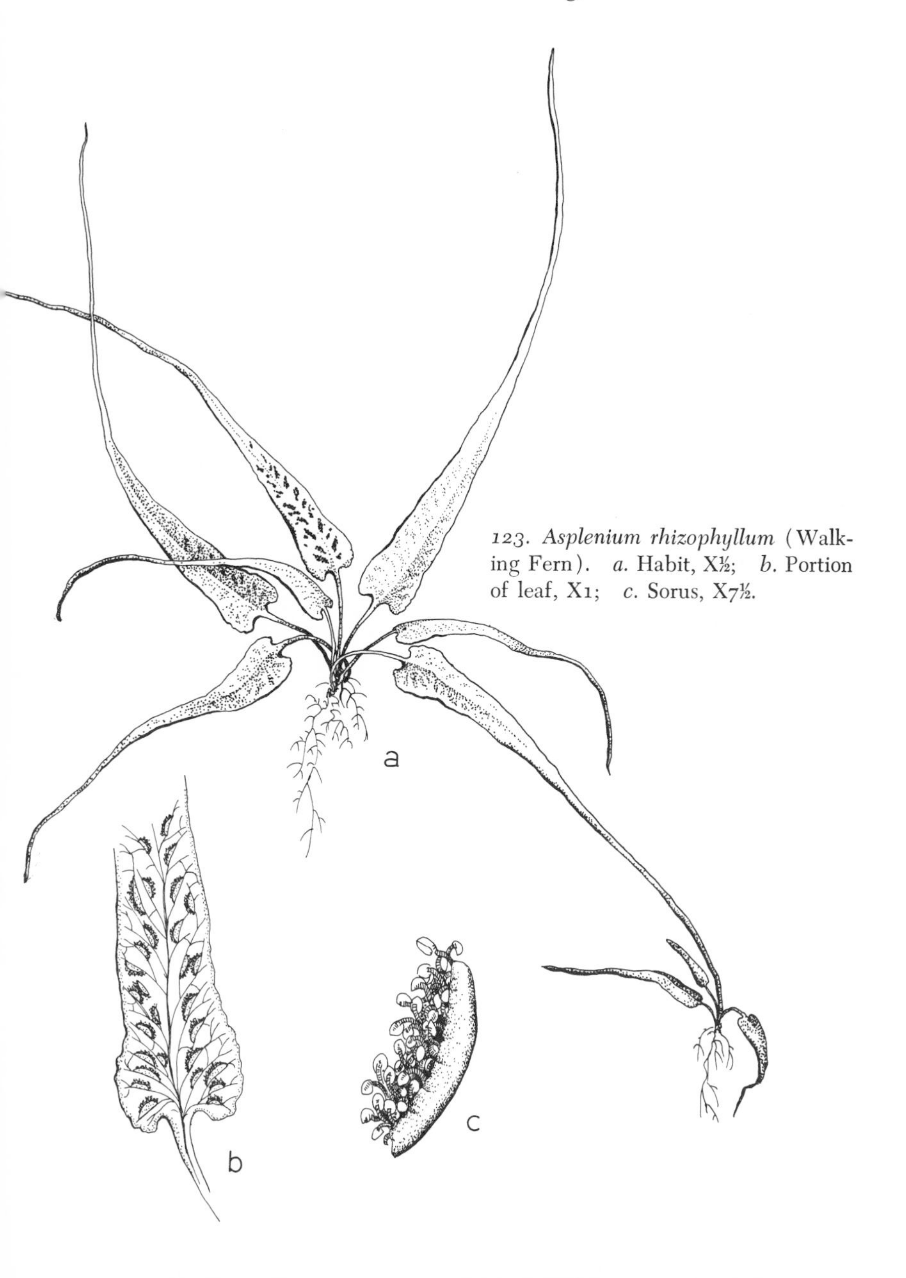

123. *Asplenium rhizophyllum* (Walking Fern). *a.* Habit, X½; *b.* Portion of leaf, X1; *c.* Sorus, X7½.

margin free; petiole dark brown and scaly at the very base, green and glabrous above; sori elongated, scattered; indusium laterally attached; 2n = 72 (Wagner, 1954).

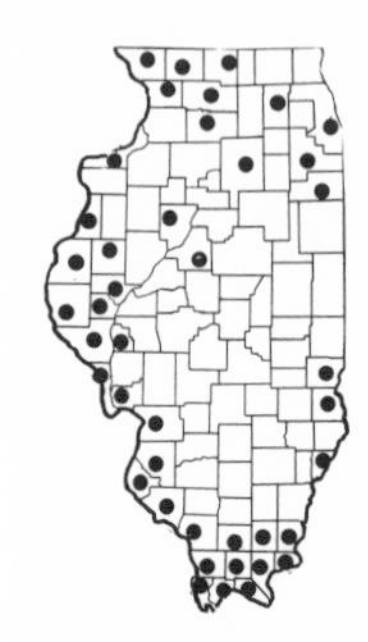

COMMON NAME: Walking Fern.
HABITAT: Rocky woodlands, either on sandstone or limestone.
RANGE: Quebec to Ontario and Minnesota, south to Oklahoma and Georgia.
ILLINOIS DISTRIBUTION: Common in the southern one-fourth of Illinois and also along the Mississippi River; absent from the central interior part of the state.

The walking fern often forms extensive, connecting colonies because of the ability to root at the elongated leaf tips. While long placed in its own genus, *Camptosorus,* it is believed that this species might best be placed in *Asplenium,* primarily because of its somewhat similar soral conditions and its ability to hybridize with various spleenworts.

Variation exists in size and shape of leaf and degree of auriculation at the leaf base.

2. Asplenium ruta-muraria L. Sp. Pl. 1079. 1753. *Fig. 124.*

Asplenium cryptolepis Fern. in Rhodora 30:41. 1928.

Asplenium ruta-muraria L. var. *cryptolepis* (Fern.) Christ ex Massey, in Claytonia 2:34. 1935.

Small evergreen from slender, obscurely scaly rhizomes; leaves rather triangular, bipinnate or pinnate-pinnatifid, to 6 cm long, with 2–4 pairs of alternate pinnae, the pinnae glabrous, subcoriaceous, the lower with up to 7 dentate divisions, the upper undivided; petiole and rachis very slender, green, glabrous; sori elongated; indusium laterally attached; 2n = 44.

COMMON NAME: Wall-rue Spleenwort.
HABITAT: Crevices of limestone cliffs.
RANGE: Vermont to Ontario, south to Arkansas (?) and Alabama.
ILLINOIS DISTRIBUTION: Very rare; known only from "s. Ill.," collected in the middle 1800's; probably extinct in Illinois.

There seems to be little justification in following Fernald in separating the American specimens as var. *cryptolepis.*

124. *Asplenium ruta-muraria* (Wall-rue Spleenwort). *a.* Habit, X1½; *b.* Leaves, X½; *c.* Pinna with sori, X6.

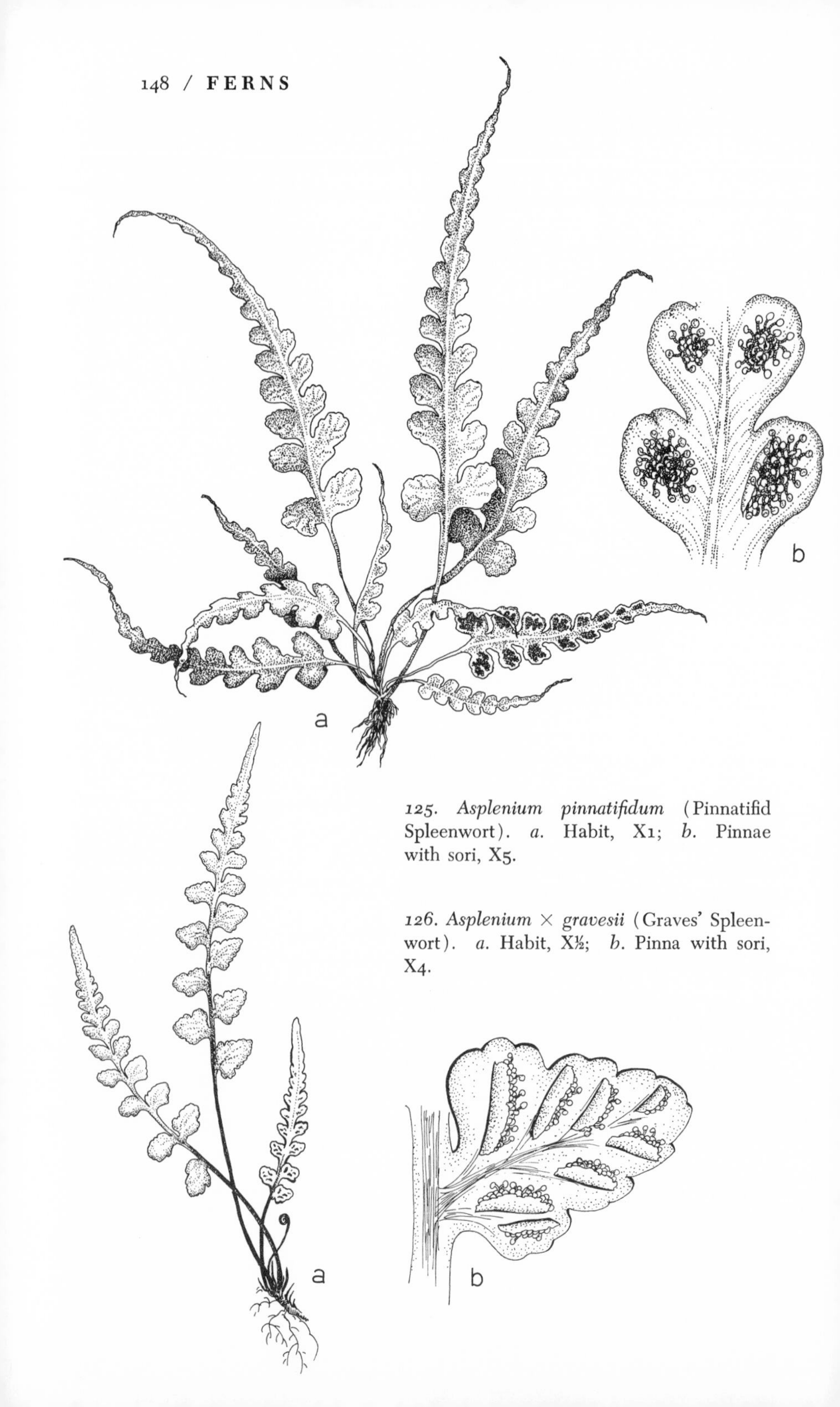

125. *Asplenium pinnatifidum* (Pinnatifid Spleenwort). *a.* Habit, X1; *b.* Pinnae with sori, X5.

126. *Asplenium* × *gravesii* (Graves' Spleenwort). *a.* Habit, X½; *b.* Pinna with sori, X4.

3. **Asplenium pinnatifidum** Nutt. Gen. 2:251. 1818. *Fig. 125.*

Evergreen from a slender, creeping rhizome; leaves pinnatifid, rarely with the lowest pair of pinnae distinct, subcoriaceous, glabrous, short- to long-tapering at the apex, the lobes entire or crenulate; petiole brown below, green above, glabrous; rachis green, glabrous; sori elongated, often confluent; indusium laterally attached; 2n = 144 (Wagner, 1954).

COMMON NAME: Pinnatifid Spleenwort.
HABITAT: Crevices of sandstone cliffs.
RANGE: New Jersey to Oklahoma, east to Georgia.
ILLINOIS DISTRIBUTION: Occasional in the southern one-fourth of the state; absent elsewhere.

This interesting species possesses a most variable blade. The degree of lobing and the degree of apex attenuation vary even on the same plant. Most specimens are pinnatifid throughout, although some specimens may occur in which the lowest pair of pinnae is distinct. If more than the lowest pair of pinnae is distinct, then one of the other species should be suspected. Forms are known in which development of the blade is highly reduced (Mohlenbrock, 1956). Hybrids such as *A.* × *gravesii* and *A.* × *trudellii,* which may resemble *A. pinnatifidum,* have abortive spores.

4. **Asplenium × gravesii** Maxon, in Am. Fern Journ. 8:1. 1918. *Fig. 126.*

Asplenium pinnatifidum var. *gravesii* (Maxon) Clute, Our Ferns 209. 1938.

Evergreen from a slender, creeping rhizome; leaves once-pinnate below, pinnatifid above, subcoriaceous, glabrous, short-tapering to the apex, the lobes crenulate; petiole brown throughout, glabrous; rachis green (rarely brown at base), glabrous; sori elongated; indusium laterally attached; spores abortive.

COMMON NAME: Graves' Spleenwort.
HABITAT: Crevices of sandstone cliffs.
RANGE: Pennsylvania, Kentucky, Illinois.
ILLINOIS DISTRIBUTION: Very rare; known only from Panther's Den, Union County (crevice of sandstone cliff, *R. R. MacMahon* in 1960 [as *A. pinnatifidum*]).

This species is a hybrid between *A. pinnatifidum* and *A. bradleyi.* It is distinguished from the former by having several pairs of distinct pinnae and by the petiole which is entirely brown; from the latter it is distinguished by the usually entirely green rachis and the lesser degree of leaf-cutting. The spores are wrinkled and abortive. Both parents occur where *A.* × *gravesii* was found.

5. **Asplenium × trudellii** Wherry, in Am. Fern Journ. 15:49. 1925. *Fig. 127.*

Asplenium pinnatifidum var. *trudellii* (Wherry) Clute, Our Ferns 209. 1938.

Evergreen from a slender, creeping rhizome; leaves once-pinnate below, pinnatifid above, subcoriaceous, glabrous, rather long-tapering to the apex, the pinnae crenulate or shallowly lobed; petiole brown at base, green above, glabrous; rachis green, glabrous; sori elongated; indusium laterally attached; spores abortive; 2n = 108 (Wagner, 1954).

COMMON NAME: Trudell's Spleenwort.

HABITAT: Crevices of sandstone cliffs.

RANGE: New Jersey to Illinois (?), south to Alabama.

ILLINOIS DISTRIBUTION: Known only from Fern Rocks (Giant City State Park), Jackson County (or Union), *G. H. French 3719* in 1871 (originally identified as *A. pinnatifidum*).

This specimen has been checked by Dr. Wagner who suggests that since *A. montanum,* one of the parents of *A.* × *trudellii,* is unknown from Illinois, French may have confused the locality. In the French collection of *Asplenium* at Southern Illinois University, there is also *French 3718,* a true specimen of *A. pinnatifidum,* from Fern Rocks, collected in 1871.

This species is a hybrid between *A. pinnatifidum* and *A. montanum.* The nearest known station of *A.* × *trudellii* to Illinois is in Franklin County, Tennessee. This hybrid should be searched for diligently in southern Illinois in an attempt to verify the French collection.

Asplenium × *trudellii* is much like *A.* × *gravesii* except that *A.* × *trudellii* has a petiole which is brown for only half its length.

127. *Asplenium* × *trudellii* (Trudell's Spleenwort). *a*. Habit, X½; *b*. Pinna with sori, X3½.

128. *Asplenium* × *kentuckiense* (Kentucky Spleenwort). X1.

6. Asplenium × kentuckiense McCoy, in Am. Fern Journ. 26:104. 1936. *Fig. 128.*

Asplenium pinnatifidum var. *kentuckiense* (McCoy) Clute, Our Ferns 209. 1938.

Tufted evergreen from short-creeping rhizomes; leaves pinnate about one-half their length, pinnatifid above, to 20 cm long, the pinnae glabrous, crenulate, petiolulate; petiole brown throughout, glabrous; rachis brown below, green above; sori elongated; indusium laterally attached; spores abortive.

COMMON NAME: Kentucky Spleenwort.

HABITAT: Crevices of sandstone cliffs.

RANGE: Ohio; Illinois; Kentucky; Arkansas.

ILLINOIS DISTRIBUTION: Very rare; only "s. Ill.," Earle; Union Co.: Pine Hills, October 9, 1965, *J. Ebinger 6075.* Anywhere from 4–6 pairs of the lowest pinnae are petiolulate.

This species is reputedly the hybrid between *A. pinnatifidum* and *A. platyneuron.* The rachis, which is brown nearly half its length, recalls the condition in *A. bradleyi,* a bipinnate species. Smith, Bryant, and Tate (1961) report 108 chromosomes without any pairing.

7. Asplenium bradleyi D. C. Eaton, in Bull. Torrey Club 4:11. 1873. *Fig. 129.*

Evergreen perennial, from slender rhizomes; leaves bipinnate to bipinnate-pinnatifid, to 30 cm long, with up to 15 pairs of pinnae, the pinnae subcoriaceous, glabrous, the ultimate segments crenulate; petiole brown throughout, glabrous; rachis brown for half its length, green above, glabrous; sori elongated; indusium attached laterally; $2n = 144$ (Wagner, 1954).

COMMON NAME: Bradley's Spleenwort.

HABITAT: Crevices of sandstone cliffs.

RANGE: New York to Oklahoma, east to Georgia.

ILLINOIS DISTRIBUTION: Rare; known from the extreme southern counties of Randolph, Jackson, and Union; first discovered in Illinois in 1954.

This species exhibits the highest degree of leaf-cutting of any species of *Asplenium* in Illinois. The rachis, which is brown for half its length, is another good distinguishing character.

129. *Asplenium bradleyi* (Bradley's Spleenwort). *a*. Habit, 1½; *b*. Habit, X½; *c*. Pinna with sori, X3.

8. Asplenium × ebenoides R. R. Scott, in Gardener's Monthly 7:267. 1865. *Fig. 130.*

Asplenium platyneuron × Camptosorus rhizophyllus Slosson, in Bull. Torrey Club 29:487. 1902.

Asplenosorus × ebenoides (R. R. Scott) Wherry, in Am. Fern Journ. 27:56. 1937.

Evergreen from a slender, creeping rhizome; leaves once-pinnate below, pinnatifid above, subcoriaceous, glabrous, narrowly caudate at the apex, the pinnae crenulate, variable in size and shape; petiole brown throughout, glabrous; rachis brown for three-fourths its length, green at the tip, glabrous; sori elongated; indusium laterally attached; 2n = 144 (Wagner, 1954).

COMMON NAME: Scott's Spleenwort.

HABITAT: Rocky, sandstone woodlands.

RANGE: Vermont to Missouri, southeast to Alabama.

ILLINOIS DISTRIBUTION: Rare; known only from Alexander, Pope, and Jackson Counties.

This species is a hybrid between *A. rhizophyllum* and *A. platyneuron*. The dark petiole recalls *A. platyneuron*, while the tapering apex of the blade recalls *A. rhizophyllum*. The degree of leaf-cutting, which is highly variable, is intermediate between the unlobed *A. rhizophyllum* and the once-pinnate *A. platyneuron*.

9. Asplenium trichomanes L. Sp. Pl. 1080. 1753. *Fig. 131.*

Tufted evergreen, from slender, scaly rhizomes; leaves once-pinnate, to 25 cm long, with up to 20 pairs of pinnae, the pinnae mostly opposite, subcoriaceous, glabrous, crenate, asymmetrical at base, without auricles; petiole and rachis black, glabrous; sori elongated; indusium attached laterally; n = 72 (Manton, 1950).

COMMON NAME: Maidenhair Spleenwort.

HABITAT: Crevices of shaded cliffs, both limestone and sandstone.

RANGE: Quebec to British Columbia, south to Oregon, Colorado, Oklahoma, and Georgia.

ILLINOIS DISTRIBUTION: Occasional in the southern one-fourth of Illinois; absent elsewhere.

This delicate fern is distinguished from both *A. platyneuron* and *A. resiliens* by the lack of auriculate pinnae.

130. *Asplenium* × *ebenoides* (Scott's Spleenwort). *a.* Habit, X½; *b.* Pinnae with sori, X2.

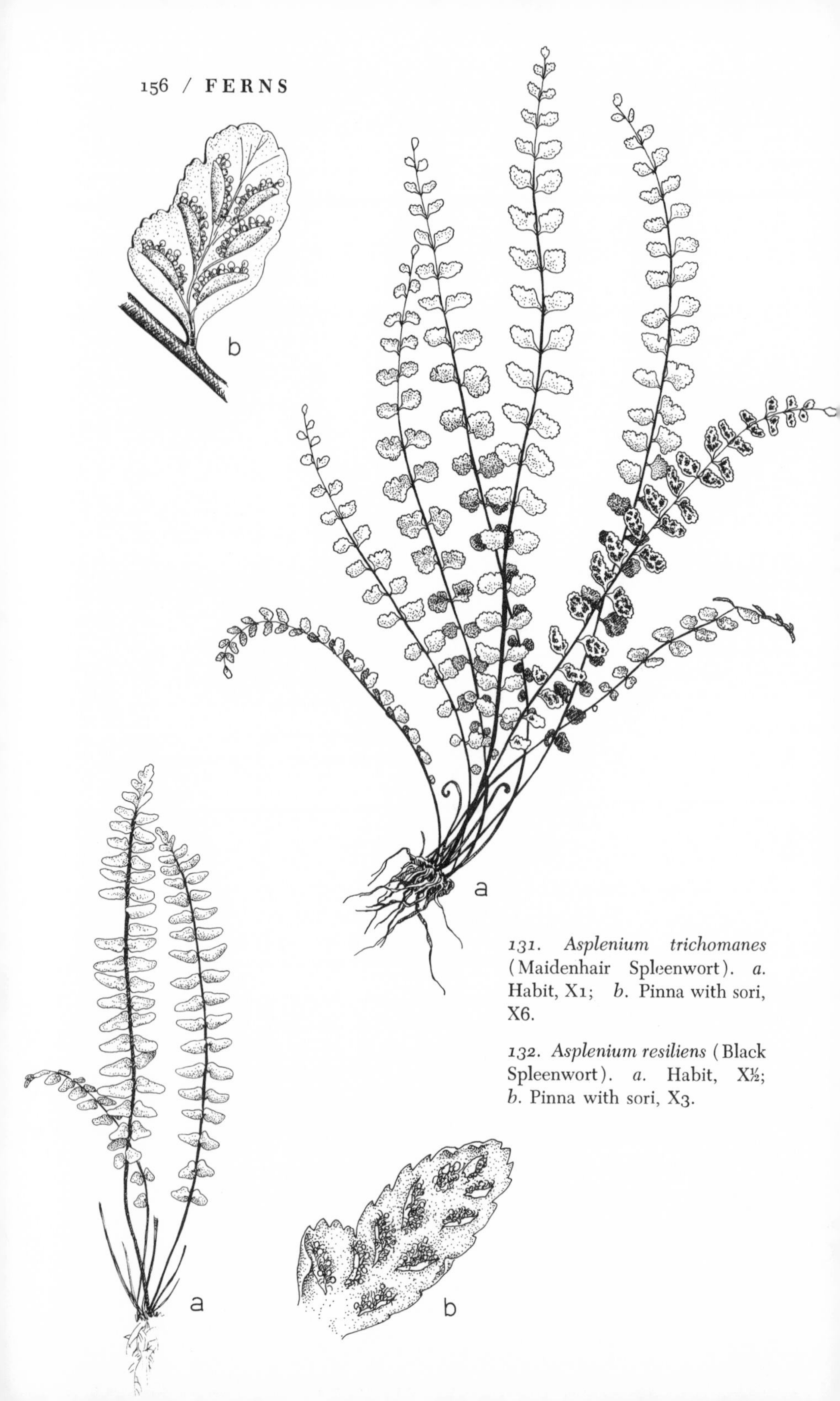

131. *Asplenium trichomanes* (Maidenhair Spleenwort). *a.* Habit, X1; *b.* Pinna with sori, X6.

132. *Asplenium resiliens* (Black Spleenwort). *a.* Habit, X½; *b.* Pinna with sori, X3.

10. Asplenium resiliens Kunze, in Linnaea 18:331. 1844. *Fig. 132.*

Asplenium parvulum Mart. & Gal. in Mem. Acad. Sci. Brux. 15:60. 1842, non Hook. (1840).

Wiry evergreen, from scaly rhizomes; leaves once-pinnate, to 30 cm long, with up to 30 pairs of pinnae, the pinnae mostly opposite, subcoriaceous, glabrous, serrulate, stalked and auriculate at base, the auricles not overlapping the rachis; petiole and rachis dark brown, glabrous; sori elongated; indusium attached laterally; 2n = 108.

COMMON NAME: Black Spleenwort.

HABITAT: Crevices of limestone cliffs.

RANGE: Pennsylvania to Kansas, south to Arizona and Florida; Central America; West Indies; South America.

ILLINOIS DISTRIBUTION: Very rare; Jackson and Union Counties.

The stalked pinnae distinguish this species from *A. platyneuron*, the other auriculate species in Illinois. *Asplenium resiliens,* because of its auriculate pinnae, is distinguished readily from *A. trichomanes*. *Asplenium resiliens* has 2n = 108 chromosomes, but has a peculiar, apogamous life cycle in which both generations have the same number.

11. Asplenium platyneuron (L.) Oakes ex D. C. Eaton, Ferns N. Am. 1:24. 1878. *Fig. 133.*

Acrostichum platyneuros L. Sp. Pl. 1069. 1753.

Asplenium ebeneum Ait. Hort. Kew. 3:462. 1789.

Wiry evergreen, from rather thick rhizomes; leaves once-pinnate, dimorphic, the sterile to 10 cm long, with up to 20 pairs of pinnae, the fertile to 50 cm long, with up to 50 pairs of pinnae, the pinnae mostly alternate, subcoriaceous, glabrous, acute, serrate, with auricles at base which overlap the rachis; petiole and rachis purple-brown, shining, glabrous; sori elongated; indusium laterally attached; 2n = 72 (Wagner, 1954).

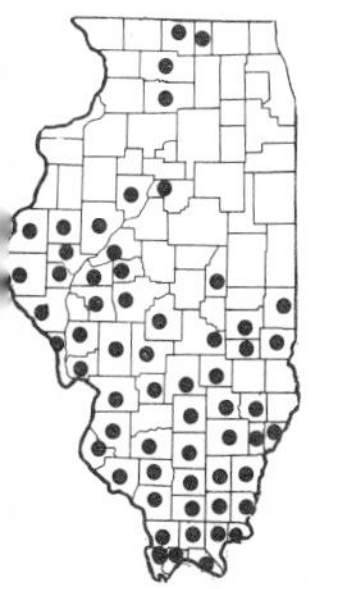

COMMON NAME: Ebony Spleenwort.

HABITAT: Dry or moist woodlands.

RANGE: Quebec to Ontario, south to Colorado, Texas, and Florida; West Indies; South America; South Africa.

ILLINOIS DISTRIBUTION: Common in the southern half of Illinois, becoming less abundant northward; apparently absent from the northeastern counties.

This is the most common species of *Asplenium* in Illinois, where it occurs in both dry and moist habitats. There is variation in degree of serration on the pinnae. The auricle at the pinna base overlaps the rachis, thus making this species easily distinguished from *A. resiliens.*

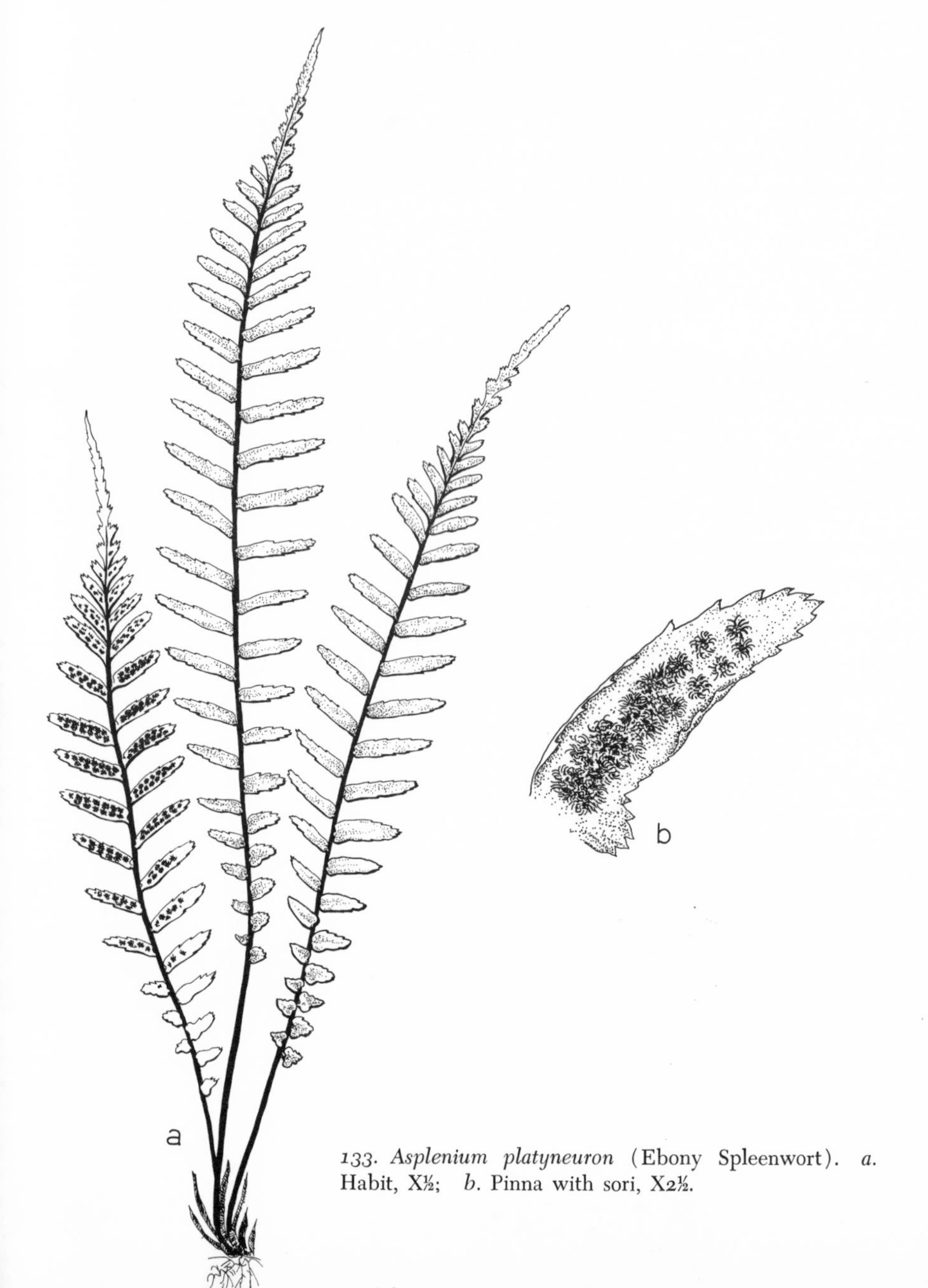

133. *Asplenium platyneuron* (Ebony Spleenwort). *a.* Habit, X½; *b.* Pinna with sori, X2½.

17. *Woodsia* R. BR. – Woodsia

Rootstock compact; leaves pinnate-pinnatifid to bipinnate-pinnatifid; sori round; indusium inferior (i.e., attached beneath the sorus).

A recent monograph of *Woodsia* has been prepared by Brown (1964).

KEY TO THE SPECIES OF Woodsia IN ILLINOIS

1. Blades bipinnate-pinnatifid; leaves generally over 25 cm long; petiole stramineous, unjointed; scales of rhizome few__1. *W. obtusa*
1. Blades pinnate-pinnatifid; leaves generally less than 20 cm long; petiole brown, jointed below the middle; scales of rhizome numerous____________________________________2. *W. ilvensis*

1. Woodsia obtusa (Spreng.) Torr. Cat. Pl. in Geol. Rep. N. Y. 195. 1840. *Fig. 134.*

Polypodium obtusum Spreng. in Anleit. Kennt. Gewächse 3:92. 1804.
Alsophila perriniana Spreng. in Nova Acta 10:232. 1821.
Woodsia perriniana (Spreng.) Hook. & Grev. Icon. Fil. Pl. 68. 1828.

Occasionally evergreen perennial from short rhizomes; leaves bipinnate to bipinnate-pinnatifid, to 35 cm long, with up to 20 pairs of pinnae, the pinnae glandular, the ultimate divisions obtusely lobed; petiole light brown, chaffy, the rachis glandular; sori round, borne on the back of the leaf segments; indusium attached beneath the sorus, splitting into 5–6 broad segments; n = 41.

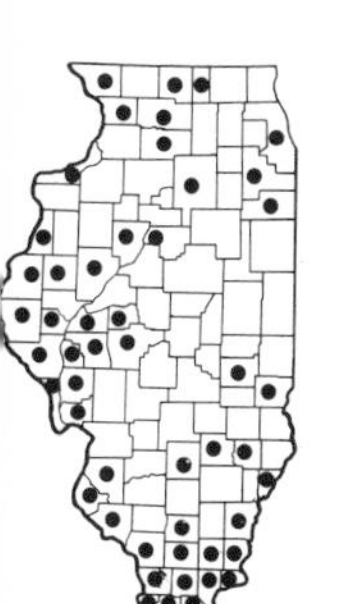

COMMON NAME: Common Woodsia.
HABITAT: Dry or moist rocky woods, occasionally on very exposed sandstone ledges.
RANGE: Nova Scotia to Minnesota, south to Texas and Florida; Alaska to British Columbia.
ILLINOIS DISTRIBUTION: Occasional to common, except in the east-central counties.
The Common Woodsia is usually deciduous, but in protected situations, the leaves may persist throughout the winter.

Cystopteris fragilis var. *protrusa* is a similar taxon, but lacks glandular blades and has glabrous or very sparsely chaffy petioles.

2. Woodsia ilvensis (L.) R. Br. Prodr. Fl. Nov. Holl. 1:158. 1810. *Fig. 135.*

Acrostichum ilvense L. Sp. Pl. 1071. 1753.

Nephrodium rufidulum Michx. Fl. Bor. Am. 2:269. 1803.

Deciduous perennial from a branched rhizome; leaves pinnate-pinnatifid to bipinnate, to 20 cm long, with up to 20 pairs of pinnae, the pinnae firm, puberulent and chaffy on the lower surface, crenately lobed; petioles dark brown below, lighter above, chaffy, jointed near base; sori round, often confluent, borne on the back of the leaf segments; indusium attached beneath the sorus, splitting into many thread-like segments; n = 41.

COMMON NAME: Rusty Woodsia.

HABITAT: Mostly dry, sandstone cliffs.

RANGE: Labrador to Alaska, south to British Columbia, Illinois, and North Carolina.

ILLINOIS DISTRIBUTION: Very rare; known only from Ogle County (sandstone cliffs, Castle Rock, near Oregon). First collected by *M. B. Waite* in 1885.

The Rusty Woodsia, a northern species, is distinguished by its jointed petiole and chaffy lower leaf surfaces.

18. Cystopteris BERNH. – Fragile Fern

Rhizome compact or elongate; leaves bipinnate to bipinnate-pinnatifid; sori round; indusium attached beneath the sorus.

A monograph of the genus has been prepared by Blasdell (1963).

KEY TO THE SPECIES OF Cystopteris IN ILLINOIS

1. All veins of ultimate leaf segments running to the sinuses ------ 1. *C. bulbifera*
1. Some or all the veins of ultimate leaf segments running to the teeth.
   2. All veins of ultimate leaf segments running to the teeth ------ 2. *C. fragilis*
   2. Some veins of ultimate leaf segments running to the teeth, some running to the sinuses.
      3. Indusium acute and toothed at the apex, eglandular ----- 2. *C. fragilis*
      3. Indusium truncate and entire at the apex, more or less glandular ------ 3. *C.* × *tennesseensis*

134. Woodsia obtusa (Common Woodsia). *a.* Habit, X½; *b.* Pinna with sori, X3½.

135. Woodsia ilvensis (Rusty Woodsia). *a.* Habit, X½; *b.* Pinna, X4.

a

b

c

136. *Cystopteris bulbifera* (Bladder Fern). *a.* Leaf, X¼; *b.* Pinna with sori, X2; *c.* Pinna with bulblet, X1¼.

1. **Cystopteris bulbifera** (L.) Bernh. in Neu. Journ. Bot. Schrad. 1 (2):10. 1806. *Fig. 136.*

Polypodium bulbiferum L. Sp. Pl. 1091. 1753.

Nephrodium bulbiferum (L.) Michx. Fl. Bor. Am. 2:268. 1803.

Filix bulbifera (L.) Underw. Nat. Ferns 119. 1900.

Deciduous perennial from a short, stout rhizome; leaves bipinnate, to 70 cm long, tapering to the tip, the pinnae membranous, to 30 pairs, usually bearing minute, glandular hairs and small bulblets on the under surface, with the veins running to the sinuses of the pinnules; petiole slender, black at base, stramineous above, glabrous; sori round, borne on the back of the leaf segments; indusium brownish, attached mostly beneath the sorus, but to one side; n = 42 (Blasdell, 1963).

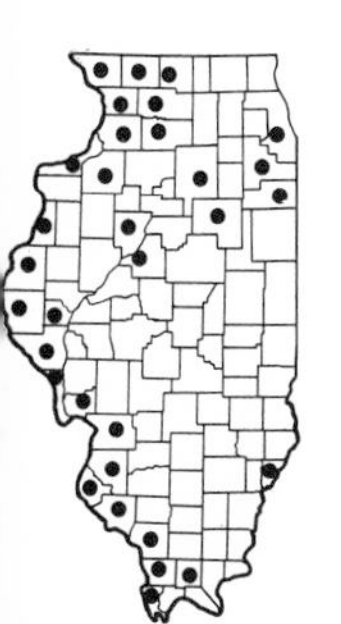

COMMON NAME: Bladder Fern.

HABITAT: Mostly limestone cliffs and woodlands.

RANGE: Newfoundland to Manitoba, south to Arizona and Georgia.

ILLINOIS DISTRIBUTION: Occasional in the counties along the Mississippi and Illinois Rivers, otherwise rare.

This fern may be recognized from *C. fragilis* in that all its veins terminate in the sinuses, rather than in the teeth. The leaves of *C. bulbifera* are generally longer, also.

The bulblets, which readily fall off and may quickly grow into young plants, are borne on the lower surface of the blades near the axils of the pinnae or pinnules. Fertile leaves tend to be longer and more long-tapering.

2. **Cystopteris fragilis** (L.) Bernh. in Neu. Journ. Bot. Schrad. 1 (2):26. 1806.

Polypodium fragile L. Sp. Pl. 1091. 1753.

Filix fragilis (L.) Gilib. Exerc. Phyt. 558. 1792.

Aspidium fragile (L.) Sw. in Journ. Bot. Schrad. 1800 (2):40. 1801.

Deciduous perennials from short, scaly rhizomes or long-creeping rhizomes; leaves bipinnatifid to bipinnate-pinnatifid, to 30 cm long, the pinnae membranous, glabrous, or with scattered glandular hairs, with the veins running to the teeth and sometimes to the sinuses of the pinnules; petiole slender, stramineous, or brown for half its length, glabrous or nearly so; sori round, borne on the back of the leaf segments; indusium whitish, 0.5–1.0 mm long, attached beneath the sorus, but to one side, acute to truncate or rounded, the tip deeply lobed or entire; spores echinate, dark brown, 27–53 μ in diameter.

Three varieties, sometimes difficult to distinguish, occur in Illinois. The treatment of *C. fragilis* follows that of Weatherby (1935). The recent treatment by Blasdell (1963) does not seem to present a realistic interpretation of the variations of this species, at least as found in Illinois.

a. All veins of ultimate leaf segments running to the teeth.
  b. Rhizome short, scaly; petiole brown for half its length; average spore size 32–53 μ ---------------- 2a. *C. fragilis* var. *fragilis*
  b. Rhizome long-creeping, densely hairy; petiole stramineous, except at the base; average spore size 27–32 μ -- 2b. *C. fragilis* var. *protrusa*
a. Some veins of ultimate leaf segments running to the teeth, some to the sinuses ------------------------- 2c. *C. fragilis* var. *mackayi*

2a. Cystopteris fragilis (L.) Bernh. var. **fragilis** *Fig. 137.* Rhizome short, scaly; all veins of ultimate leaf segment running to the teeth; petiole brown for half its length; indusium at least 1 mm long, ovate, acute, toothed at the apex; spores averaging 32–53 μ in diameter; n = 84, 126, 168 (Blasdell, 1963).

COMMON NAME: Fragile Fern.

HABITAT: Moist woodlands.

RANGE: Newfoundland to Alaska, south to California, Texas, and Virginia.

ILLINOIS DISTRIBUTION: Rare; known only from Union County (damp woods, Pine Hills Recreation Area, *J. Hinners* in 1962).

The acute, ovate indusium and all the veins running to the teeth distinguish this variety from var. *mackayi*, while the short, scaly rhizome and slightly larger spore distinguish it from var. *protrusa*.

137. *Cystopteris fragilis* var. *fragilis* (Fragile Fern). *a.* Leaf, X½; *b.* Pinna, X4½; *c.* Sorus, X20.

138. *Cystopteris fragilis* var. *protrusa* (Fragile Fern). *a*. Habit, X½; *b*. Pinna with sori, X4½; *c*. Sorus, X20.

139. *Cystopteris fragilis* var. *mackayi* (Fragile Fern). *a*. Habit, X½; *b*. Sorus, X4½.

2b. Cystopteris fragilis (L.) Bernh. var. **protrusa** Weatherby, in Rhodora 37:373. 1935. *Fig. 138.*

Cystopteris protrusa (Weatherby) Blasdell, in Mem. Torrey Club 21 (4):41. 1963.

Rhizome elongate, creeping; blade to 22 cm long, with scattered glandular hairs, the veins running to the marginal teeth; petiole stramineous; indusium about 0.5 mm long, entire or shallowly divided at the apex; spores echinate, brown, 27–32 μ in diameter; n = 42 (Blasdell, 1963).

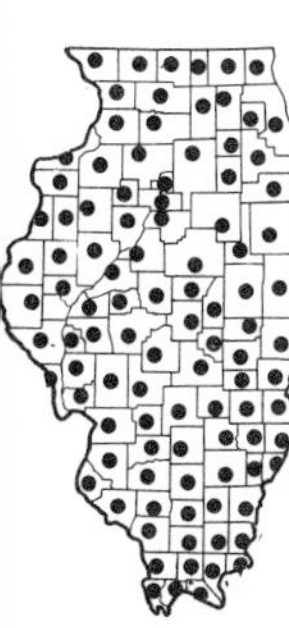

COMMON NAME: Fragile Fern.

HABITAT: Moist woodlands.

RANGE: New York to Minnesota, south to Oklahoma and Georgia.

ILLINOIS DISTRIBUTION: Common; in every county.

This fern is probably the most common fern throughout the entire state. Although Blasdell (1963) proposes to treat this as a species, it appears to be considered better as a variety of *C. fragilis.*

2c. Cystopteris fragilis (L.) Bernh. var. **mackayi** Laws. Fern Flora Can. 233. 1889. *Fig. 139.*

Rhizome short, scaly; some veins of ultimate leaf segment running to the teeth, some to the sinuses; petiole brown for half its length; indusium about 0.5 mm long, entire or shallowly toothed at the rounded or truncate apex; spores averaging 32–53 μ in diameter; tetraploid.

COMMON NAME: Fragile Fern.

HABITAT: Moist woodlands.

RANGE: Quebec to South Dakota, south to Missouri and North Carolina.

ILLINOIS DISTRIBUTION: Apparently rare; only known from LaSalle County.

This is a very rare variety of *C. fragilis,* thus far known only from a single collection. The tiny indusium is like that of *C. fragilis* var. *protrusa,* but the short rhizome relates this variety to *C. fragilis* var. *fragilis.*

Blasdell (1963) considers *C. fragilis* var. *mackayi* to represent an introgressant hybrid between *C. diaphana* (Bory) Blasdell and *C. fragilis.*

3. **Cystopteris × tennesseensis** Shaver, in Journ. Tenn. Acad. Sci. 25 (2):107. 1950. *Fig. 140.*

Cystopteris fragilis f. *simulans* Weatherby, in Rhodora 37:376. 1935.

Cystopteris fragilis var. *tennesseensis* (Shaver) McGregor, in Am. Fern Journ. 40:202. 1950.

Rhizome rather short, scaly; blades bi- to tripinnatifid, to 31 cm long, with scattered glandular hairs, the obtuse pinnules obtusely lobed and toothed, the veins running to the teeth and to the sinuses; indusium cup-shaped, truncate, more or less glandular; spores dark brown, echinate, 32–42 μ in diameter; n = 84 (Blasdell, 1963).

COMMON NAME: Tennessee Fragile Fern.

HABITAT: Moist woodlands.

RANGE: Maryland to Kansas, south to Oklahoma and North Carolina.

ILLINOIS DISTRIBUTION: Not common; known from Champaign and Will counties.

Blasdell (1963) equates *C. fragilis* f. *simulans* with *C. tennesseensis.* This is a putative hybrid between *C. fragilis* var. *protrusa* and *C. bulbifera.* It shows the venation characters of both parents.

Excluded Species

A specimen of the cultivated *Pteris multifida* Poir. was collected by J. Schneck from a well in Mt. Carmel on April 15, 1902. It has been excluded in the enumeration of Illinois ferns.

140. Cystopteris × *tennesseensis* (Tennessee Fragile Fern). *a.* Habit, X½; *b.* Pinna with sori, X6; *c.* Sorus, X17½.

Order Salviniales

This order is composed of two families in Illinois, each with a single genus. Both genera are aquatic and produce both megaspores and microspores enclosed in sporangia aggregated together to comprise the sporocarp.

KEY TO THE FAMILIES OF **Salviniales** IN ILLINOIS

1. Plants rooted; blades shaped like a 4-leaf clover; blades and sporocarps with long stalks______________________Marsileaceae
1. Plants floating; blades very small, bilobed; blades and sporocarps sessile__________________________________Salviniaceae

MARSILEACEÆ – WATERCLOVER FAMILY

Only the following genus occurs in Illinois.

1. *Marsilea* L. – Waterclover

Aquatic from rooted rhizomes; blades 4-parted, long-stipitate; sporocarps stalked.

Only the following species has been found in Illinois.

1. Marsilea quadrifolia L. Sp. Pl. 1099. 1753. *Fig. 141.*

Blades glabrous or nearly so, 4-parted, up to 20 mm across, the petioles 15–30 cm long; sporocarps ellipsoid, 4–5 mm long, 3–4 mm thick, glabrous, punctate, long-stalked, the stalk arising from near the base of the petiole.

COMMON NAME: Waterclover.
HABITAT: Ponds or lakes.
RANGE: Native to Europe; introduced, frequently from fish tanks, in several of the United States.
ILLINOIS DISTRIBUTION: Scattered.

141. *Marsilea quadrifolia* (Waterclover). *a.* Habit, X½. *b.* Leaf, X1.

SALVINIACEÆ – WATER FERN FAMILY

Only the following genus occurs in Illinois.

1. *Azolla* LAM. – Mosquito Fern

Small floating aquatics; stems branched, bearing rootlets; leaves very small, the blades bilobed; sporocarps sessile, borne on the slender stems.

The latest treatment of the New World species of *Azolla* is by Svenson (1944).

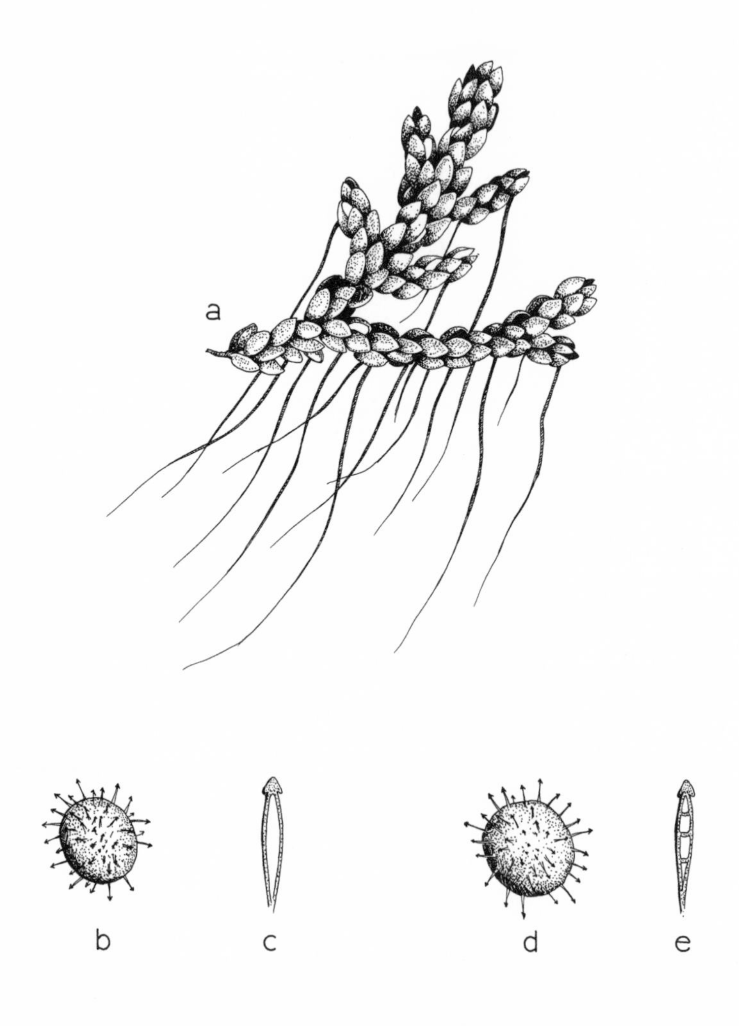

142. *Azolla caroliniana* (Mosquito Fern)—*a*. Habit, X7½; *b*. Microsporocarp, X500; *c*. Glochidium, X500; *Azolla mexicana* —*d*. Microsporocarp, X500; *e*. Glochidium, X500.

KEY TO THE SPECIES OF **Azolla** IN ILLINOIS

1. Plants less than 1 cm in diameter; leaves about 0.5 mm long; glochidia without cross-walls________________1. *A. caroliniana*
1. Plants 1 cm in diameter or larger; leaves about 0.7 mm long or longer; glochidia with cross-walls_______________2. *A. mexicana*

1. **Azolla caroliniana** Willd. Sp. Pl. 5:541. 1810. *Fig. 142a, b, c.*

Plants 5–10 mm in diameter; upper lobe of leaf about 0.5 mm long; glochidia without cross-walls; megaspores unknown.

COMMON NAME: Mosquito Fern.
HABITAT: Standing water.
RANGE: Massachusetts and New York to Louisiana; Cuba; Puerto Rico; Jamaica; Brazil (?), according to Svenson (1944).
ILLINOIS DISTRIBUTION: Very rare; positively known only from St. Clair County (in sink, 3 miles east of Dupo, October 1, 1939, *A. J. Seibert 1134*).

This species is only definitely distinguished from *A. mexicana* by its lack of cross-walls in the glochidia. Megaspores have never been found for *A. caroliniana*. Since most Illinois material is sterile, the size of the plant has to be relied upon in most cases. The only specimen from Illinois admitted to this species is one bearing microsporocarps from St. Clair County. Godfrey, Reinert, and Houk (1961) have questioned the validity of presence or absence of cross-walls in the glochidia for species determination.

2. **Azolla mexicana** Presl, in Abh. Bohm. Ges. Wiss. V. 3:150. 1845. *Fig. 142d, e.*

Plants 10–15 mm in diameter; upper lobe of leaf about 0.7 mm long; glochidia with cross-walls; megaspores 0.4–0.5 mm in diameter, pitted near base.

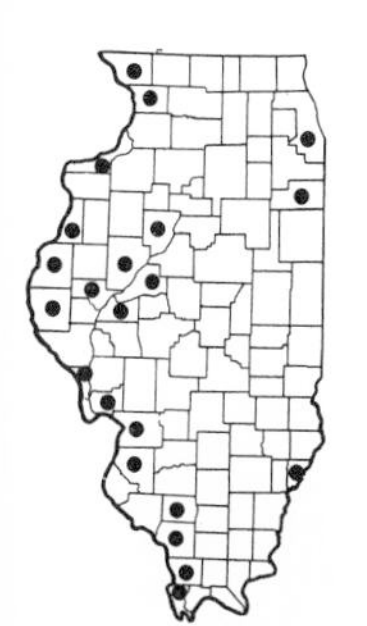

COMMON NAME: Mosquito Fern.
HABITAT: Ponds and standing water.
RANGE: Wisconsin; Illinois; Missouri; British Columbia south to California, east to Utah; Mexico; Central America; South America.
ILLINOIS DISTRIBUTION: Local, scattered throughout the state.

Several specimens of this species from Illinois are in the fruiting stage.

Summary of the Taxa Treated in This Volume

Orders and Families	*Genera*	*Species*	*Varieties*
Order 1. **Lycopodiales**			
Family 1. Lycopodiaceae	1	6	1
Order 2. **Selaginellales**			
Family 2. Selaginellaceae	1	2	
Order 3. **Isoetales**			
Family 3. Isoetaceae	1	3	
Order 4. **Equisetales**			
Family 4. Equisetaceae	1	11	
Order 5. **Ophioglossales**			
Family 5. Ophioglossaceae	2	6	1
Order 6. **Filicales**			
Family 6. Hymenophyllaceae	1	1	
Family 7. Osmundaceae	1	3	
Family 8. Polypodiaceae	18	46	4
Order 7. **Salviniales**			
Family 9. Marsileaceae	1	1	
Family 10. Salviniaceae	1	2	
Totals	28	81	6

GLOSSARY

LITERATURE CITED

INDEX OF PLANT NAMES

GLOSSARY

Acuminate. Gradually tapering to a point.

Anastomosing. Forming a network of cross-veins.

Annual. Living only for one year.

Annulus. A ring of cells of the sporangium which functions in spore dispersal.

Apical. Relating to the apex or tip.

Apiculate. Ending abruptly in a small sharp tip.

Appressed. Lying flat or against a stalk or margin.

Areole. A small area formed by the interlocking of veins.

Asymmetrical. Of different shape on the two sides.

Attenuate. Long-tapering.

Auricle. An ear-like process.

Auriculate. Bearing an ear-like process.

Axil. The angle between the base of a leaf and the axis (stem) from which the leaf arises.

Axillary. Formed in the axils.

Bilobed. Divided into two rounded divisions.

Bipinnate. Divided once into distinct segments, with each segment in turn divided into distinct segments.

Bipinnate-pinnatifid. Divided once into distinct segments, with each segment in turn divided into distinct segments, which in turn are only partially divided.

Bipinnatifid. Divided part way to the center, with each lobe again divided part way to its center.

Bivalved. With two valves, or coverings.

Blade. The expanded portion of the leaf.

Bulblet. A small bulb, or mass of tissue capable of developing into a new plant.

Caudate. Having a slender tail-like appendage.

Chaffy. Covered with scales.

Cilia. Marginal hairs.

Ciliate. Bearing marginal hairs.

Ciliolate. Bearing very short marginal hairs.

Circinate. Coiled, referring to the young frond before expansion.

Compound. Said of a structure which is divided into distinct units.

Cone. An aggregation of sporangia and their subtending sporophylls; a strobilus.

Confluent. Running together, in reference to some sori which merge with one another at maturity.

Connate. United, when referring to like parts.

Cordate. Heart-shaped.

Coriaceous. Leathery.

Corm. A short, hard or fleshy, upright underground stem.

Crenate. With round teeth.

Crenulate. With small, round teeth.

Cuneate. Tapering at the base.

Deciduous. Falling away, particular of leaves which drop off or wither during the winter.

Deltoid. Broadly triangular.

Dentate. With sharp teeth, the tips of which project outward.

Denticulate. With small, sharp teeth, the tips of which project outward.

Dichotomous. Forked; two-branched.

Dimorphic. Having two forms.

Echinate. Having prickles.

Eglandular. Without glands.

Elliptic. Broadest at middle, gradually tapering equally to both ends.

Elliptic-ovate. Broadest midway between middle and base, tapering gradually to apex.

Entire. Without lobes or teeth, in reference to the margins of structures.

Epiphytic. Growing upon plants.

Fertile. Bearing reproductive parts.

Flabellate. Fan-shaped.

Flaccid. Weak; flabby.

Flexible. Able to be bent readily.

Frond. The leaf of a fern, including both blade and stipe.

Gamete. A sex cell; in the male, it is the sperm; in the female, it is the egg.

Gametophyte. That plant or generation which produces gametes.

Gemmae (sing., **gemma**). Bulblets or buds capable of developing into new plants.

Glabrous. Smooth; without hairs, scales, or glands.

Gland. An enlarged, spherical body functioning as a secretory organ.

Glandular. Bearing glands.

Globoid. Referring to a solid body which is round.

Globular. Roundish.

Glochidia (sing., **glochidium**). Processes bearing barbs, as in *Azolla.*

Herbaceous. Dying back at the end of a growing season; opposite to woody.

Indusium. An outgrowth of the blade which covers or partially covers the sori.

Inferior. Below, referring to indusia attached beneath the sori.

Lanate. Woolly.

Lateral. From the side, in reference to the point of attachment of certain indusia.

Ligule. An elongated, flaplike process found on the leaf of *Isoetes* and *Selaginella.*

Linear. Elongated and uniform in width throughout.

Linear-lanceolate. Somewhat broadened at base, very gradually tapering to the apex.

Lobed. Divided into rounded segments.

Locular. Bearing cavities.

Margin. Edge, referring to the blade.

Megaspore. A spore produced in a megasporangium; a

spore which produces a female gametophyte.

Membranous. Like a membrane; thin.

Microspore. A spore produced in a microsporangium; a spore which produces a male gametophyte.

Microsporocarp. A compound structure containing the microsporangia in *Marsilea* and *Azolla*.

Mycorhizal. Pertaining to the association of certain fungi to roots.

Node. That place on the stem from which leaves and branchlets arise.

Oblanceolate. Opposite of lanceolate, i.e., broadest at apex, gradually tapering toward base.

Oblong. With nearly uniform width throughout, but broader than linear.

Oblongoid. Referring to a solid object which, in side view, is nearly the same width throughout, but broader than linear.

Obtuse. Rounded; blunt.

Once-pinnate. Divided once into distinct entire segments on either side of an axis.

Opaque. Not transmitting light; opposite of translucent.

Orbicular. Round.

Ovate. Broadly rounded at base, becoming narrowed above; broader than lanceolate.

Panicles. Compound group of reproductive structures.

Papillate. Bearing small warts, or papillae.

Papillose. Warty.

Peduncle. A stalk to a reproductive organ.

Pedunculate. Bearing a peduncle, or stalk.

Peltate. Supported from the lower surface by a central stalk.

Perennial. Living more than one year.

Persistent. Leaves through winter.

Petiolate. With a petiole, or leaf stalk.

Petiole. The stalk or stipe of the leaf.

Petiolule. A small stalk, referring to the stalk of a pinna.

Pinna. A primary division of a compound blade.

Pinnate-pinnatifid. Divided once into distinct segments, with each segment further partially divided.

Pinnatifid. Said of a simple leaf or leaf-part which is cleft or lobed only part way to its axis.

Pinnule. The secondary segments of a compound blade.

Prostrate. Lying flat.

Puberulent. Minutely pubescent.

Pubescent. Hairy; bearing some type of hair.

Punctate. Dotted.

Pustular. Having raised surfaces like blisters.

Rachis. That portion of the leaf which is a continuation of the petiole.

Ranked. Referring to the number of planes in which structures are borne.

Receptacle. The vascular and associated tissue that bears the sporangia.

Reflexed. Turned back.

Reniform. Kidney-shaped.

Reticulate. Resembling a network.

Revolute. Rolled under at the margin.

Rhizome. A horizontal, underground stem.

Rootlet. A small root.

Rootstock. The underground stem, usually a rhizome.

Rudiment. A trace; a remnant.

Scaly. Bearing scales, or minute epidermal outgrowths.

Scurfy. With scaly outgrowths of the epidermis.

Septate. With cross-walls.

Serrate. With sharp teeth, the tips of which project forward.

Serrulate. With very small teeth, the tips of which project forward.

Sessile. Without a stalk.

Sheath. A more or less membranous structure encircling the base of a leaf.

Simple. Said of leaves which are not divided into distinct segments; as opposed to compound.

Sinus. The cleft between two lobes or teeth.

Sorus (pl., **sori**). An aggregation of sporangia.

Spike. An elongated arrangement of reproductive structures in which the component parts are sessile.

Spinulose. With small spines or bristle-tips.

Spiny-toothed. Bearing teeth with short, sharp-pointed projections.

Sporangium. A structure bearing spores.

Spore. That structure formed within a sporangium which gives rise to the gametophyte generation.

Sporocarp. A compound structure containing the sporangia in *Marsilea* and *Azolla*.

Sporophyll. A structure subtending the sporangia; in origin, it is a modified leaf.

Sporophyte. That plant or generation which produces spores.

Sterile. Not bearing any reproductive parts.

Stipe. The stalk or petiole of the leaf.

Stipitate. Bearing a stalk.

Stomate. An opening in the epidermis of the leaf.

Stramineous. Straw-colored.

Striate. Marked with grooves, such as the aerial stem of *Equisetum*.

Subacute. Nearly short-pointed.

Subcoriaceous. Nearly leathery.

Subgloboid. Nearly round, in reference to solid objects.

Subsessile. Nearly sessile; with a very short stalk.

Subulate. With a very short, narrow point.

Taxa (sing., **taxon**). Plants or groups of plants of any taxonomic rank.

Ternate. Divided into three principal parts.

Translucent. Partly transparent.

Tripinnate. Divided three times into distinct segments.

Tubercle. A small, rounded projection; a wart.

Tuberculate. Bearing small, rounded projections.

Ultimate. The last of a series of divisions.

Undulate. Wavy, referring to the margin of a blade.

Unilocular. With one locule, or cavity.

Venation. The pattern of the veins.

Villous. With long, soft, slender, unmatted hairs.

Whorl. A circle of three or more structures arising from a node.

Winged. Bearing an expanded portion on either side of an axis.

LITERATURE CITED

Blasdell, R. 1963. A Monographic Study of the Fern Genus *Cystopteris*. Memoirs of the Torrey Botanical Club 21 (4):1–102.

Brendel, F. 1887. Flora Peoriana. Peoria, Illinois.

Britton, D. M. 1953. Chromosome Studies on Ferns. The American Journal of Botany 40:575–83.

Broun, M. 1938. Index to North American Ferns. Orleans, Massachusetts.

Brown, D. F. M. 1964. A Monographic Study of the Fern Genus *Woodsia*. Nova Hedwigia, vol. 16.

Butters, F. K. 1917. Taxonomic and Geographic Studies in North American Ferns. I. The Genus *Athyrium* and the North American Ferns Allied to *Athyrium filix-femina*. Rhodora 19:170–207.

Clausen, R. T. 1938. A Monograph of Ophioglossaceae. Memoirs of the Torrey Botanical Club 19(2):1–177.

Dowell, P. 1908. New Ferns Described as Hybrids in the Genus *Dryopteris*. Bulletin of the Torrey Botanical Club 35:135–40.

Evers, R. A. 1961. The Filmy Fern in Illinois. Biological Notes No. 44. Natural History Survey Division, Urbana, Illinois.

Fernald, M. L. 1922. *Polypodium virginianum* and *P. vulgare*. Rhodora 24:125–42.

———. 1950. Gray's Manual of Botany. 8th ed. New York: The American Book Company.

Gleason, H. A. 1952. The New Britton and Brown Illustrated Flora of the Northeastern United States and Adjacent Canada. I. New York: The New York Botanical Garden.

Godfrey, R. K., G. W. Reinert, and R. D. Houk. 1961. Observations on Microsporocarpic Material of *Azolla caroliniana*. The American Fern Journal 51:89–92.

Hauke, R. L. 1962. A Resumé of the Taxonomic Reorganization of *Equisetum*, Subgenus Hippochaete. II–IV. The American Fern Journal 52:29–35, 57–63, 123–29.

Lloyd, F. E. and L. M. Underwood. 1900. A review of the Species of *Lycopodium* of North America. Bulletin of the Torrey Botanical Club 27:147–68.

Manton, I. 1950. Problems of Cytology and Evolution in the Pteridophyta. I. London and New York: Cambridge University Press. 316 pp.

Mead, S. B. 1846. Catalogue of Plants Growing Spontaneously in the State of Illinois. Prairie Farmer 6:35–36, 60, 93, 119–22.

Mohlenbrock, R. H. 1956. An Unusual Form of *Asplenium pinnatifidum* Nutt. The American Fern Journal 46:91–93.

——— and J. W. Voigt. 1959. A Study of the Filmy Fern, *Trichomanes boschianum*. The American Fern Journal 49:76–85.

Patterson, H. N. 1876. Catalogue of the Phaenogamous and Vascular Cryptogamous Plants of Illinois. Oquawka, Illinois.

Pfeiffer, N. 1922. Monograph of the Isoetaceae. Annals of the Missouri Botanical Garden 9:79–232.

Prantl, K. 1883. Systematische Uebersicht der Ophioglosseen. Ber. Deutsch. Bot. Ges. 1:348–53.

———. 1884. Beitrage zur Systematik der Ophioglosseen. Jahrb. Bot. Gart. Berlin 3:297–350.

Schaffner, R. 1926. *Equisetum variegatum Nelsoni* a Good Species. The American Fern Journal 16:45–48.

Smith, D. M., T. R. Bryant, and D. E. Tate. 1961. New Evidence on the Hybrid Nature of *Asplenium kentuckiense*. Brittonia 13(3): 289–92.

Svenson, H. K. 1944. The New World Species of *Azolla*. The American Fern Journal 34:69–84.

Tryon, A. 1957. A Revision of the Fern Genus *Pellaea* section Pellaea. Annals of the Missouri Botanical Garden 44:125–93.

Tryon, R. M. 1941. A Revision of the Genus *Pteridium*. Rhodora 43: 1–31, 37–67.

———. 1955. *Selaginella rupestris* and its Allies. Annals of the Missouri Botanical Garden 42:1–99.

———. 1960. A Review of the Genus *Dennstaedtia* in America. Contributions from the Gray Herbarium 187:23–52. Illustrated.

Wagner, W. H. 1954. Reticulate Evolution in the Appalachian *Aspleniums*. Evolution 8:103–18.

———. 1961. Nomenclature and Typification of Two *Botrychiums* of the Southeastern United States. Taxon 10(6):165–69.

Walker, S. 1959. Cytotaxonomic Studies of Some American Species of *Dryopteris*. The American Fern Journal 49:104–12.

Weatherby, C. A. 1935. A New Variety of *Cystopteris fragilis* and Some Old Ones. Rhodora 37:373–78.

———. 1936. A List of Varieties and Forms of the Ferns of Eastern North America. The American Fern Journal 26:130–36.

Wherry, E. T. 1961. The Fern Guide. Garden City, New York: Doubleday & Company, Inc. 318 pp.

Wilce, J. H. 1965. The Section Complanata of the Genus *Lycopodium*. Nova Hedwigia, vol. 19. Weinheim, Germany.

Wilson, L. R. 1932. The Identity of *Lycopodium porophilum*. Rhodora 34:169–72.

Winterringer, G. S. and R. A. Evers. 1960. New Records for Illinois Vascular Plants. Scientific Papers Series. XI. The Illinois State Museum, Springfield.

REFERENCES EXCLUSIVELY DEVOTED TO ILLINOIS FERNS

Clute, W. N. 1904. A New Species of *Equisetum*. Fern Bulletin 12: 20–23.

Engh, J. H. and R. H. Mohlenbrock. 1964. The Ferns and Fern Allies of the Pine Hills Field Station and Environs. The American Fern Journal 54:25–38.

Evers, R. A. 1961. The Filmy Fern in Illinois. Biological Notes No. 44. Natural History Survey Division. Urbana, Illinois.

———. 1964. Illinois Flora: Notes on *Leptochloa* and *Lycopodium*. Rhodora 66:165–67.

Fell, E. W. and G. B. Fell. 1949. Ferns of Rock River Valley in Illinois. Transactions of the Illinois Academy of Sciences 42:56–62.

Hill, E. J. 1899. The Habitats of the Pellaeas. Bulletin of the Torrey Botanical Club 26:303–11.

———. 1901. The Rock Relations of the Walking Fern. Fern Bulletin 9:55.

———. 1902. The Earliest Fern. Fern Bulletin 10:78.

———. 1902. *Pellaea atropurpurea*, an Evergreen. Fern Bulletin 10:82.

———. 1905. *Equisetum scirpoides* in Illinois. Fern Bulletin 13:21–23.

———. 1910. Fern Notes. Fern Bulletin 18:65–76.

———. 1912. The Rock Relations of the Cliff-brakes. Fern Bulletin 20:1–5.

———. 1912. The Fern Flora of Illinois. Fern Bulletin 20:33–43, 73.

Jones, G. N. 1947. An Enumeration of Illinois Pteridophyta. The American Midland Naturalist 38:76–126.

Mohlenbrock, R. H. 1955. The Pteridophytes of Jackson County, Illinois. The American Fern Journal 45:143–50.

———. 1956. The Pteridophytes of Jackson County, Illinois. The American Fern Journal 46:15–22.

———. 1956. An Unusual Form of *Asplenium pinnatifidum* Nutt. The American Fern Journal 46:91–93.

———. 1958. *Dryopteris clintoniana* in Illinois. The American Fern Journal 48:122–23.

———. 1960. *Isoetes melanopoda* in Southern Illinois. The American Fern Journal 50:181–84.

Myers, R. M. 1950. A New Station for *Marsilea quadrifolia* in Illinois. The American Fern Journal 40:256.

Neill, J. 1950. *Isoetes melanopoda* Still Grows in Illinois. The American Midland Naturalist 44:251.

Palmer, E. J. 1932. Notes on *Ophioglossum engelmanni*. The American Fern Journal 22:43–47.

Pickett, F. L. 1914. A Peculiar Form of *Pellaea atropurpurea*. The American Fern Journal 4:97–101.

———. 1917. Is *Pellaea glabella* a Distinct Species? The American Fern Journal 7:3–5.

Rapp, W. F. 1946. A Further Note on *Equisetum laevigatum* var. *proliferum*. The American Fern Journal 36:19.

Reed, P. O. 1941. A Taxonomic Analysis of the Fern Flora of Illinois. Unpublished thesis. University of Illinois.

Schneck, J. 1900. *Pteris cretica* in Illinois. Botanical Gazette 29:201.

Steagall, M. M. 1927. Some Illinois Ozark Ferns in Relation to Soil Acidity. Transactions of the Illinois Academy of Sciences 19: 113–36.

Weber, W. R. and R. H. Mohlenbrock. 1958. An Unusual Form of *Asplenium bradleyi*. The American Fern Journal 48:159–61.

INDEX OF PLANT NAMES